Business Plans - How To Prepare and Implement Them

Joe Erfurt

AuthorHouse™ UK Ltd.
500 Avebury Boulevard
Central Milton Keynes, MK9 2BE
www.authorhouse.co.uk
Phone: 08001974150

First published by AuthorHouse 1/23/2008

ISBN: 978-1-4343-2109-1 (sc)

Printed in the United States of America
Bloomington, Indiana

This book is printed on acid-free paper.

The Author

Born in a South Yorkshire mining village, Joe Erfurt initially worked for British Steel and then in automation engineering in the glass container industry. This led to ten years work in Johannesburg, South Africa during which time he became a leading designer of advanced automation solutions for the glass container industry. On his return to the UK in 1983 he joined British GEC, where he began his sales career as a regional export sales manager. Initial successes in China and the USA eventually led to a general management position within America where he lived for 5 years since when he has worked for a number of companies including Siemens UK, and the American and European GEC companies.

During his period with Siemens and GE, Joe had unparalleled success in securing and managing multi-million pound contracts for the Middle East via world wide consultants and contractors, mainly Japan, France and Italy. These three being amongst the 38 countries he has travelled to on business throughout his career.

He currently runs a guest house business with his wife Diane. He shares this activity with work as a chartered marketing consultant for marketing planning and its impact upon corporate strategy. He is a fellow of the Institute of Sales and Marketing Managers a member of the Chartered Institute of Marketing and holds an MA from Portsmouth University and a Diploma in Business Management.

Front cover design by; Manuel Rodrigues

BUSINESS PLANS – HOW TO PREPARE AND IMPLEMENT THEM

INTRODUCTION

Our world of commerce consists of many different types of business activities. These same businesses may exist through product or service offerings or a mix of both depending upon market wants, needs and technology growth.
Any assessment of a companies marketing mix should give rise to a simple reason why *business plans* must be produced.

Irrespective of the business under consideration it must be seen to have the ability to both survive and to grow within the business environment it operates. This statement can be defined further as follows:-

- Business Growth (or Creation): To maximise opportunities and to minimise risks.

- Business Survival: To project accurate operational and long term strategic cash flow predictions.

Little wonder why our financial institutions mandate the production of a *business plan* before any financial support (if required) is provided? This being so, in what way should a plan be produced and where do we start?

Under normal circumstances a specific marketing plan would form the basis of a "*business plan*" planning strategy. However, marketing plans will be seen to vary depending upon the author under review. Here there is a need to consider carefully and to guard against the marketing elucidation techniques or otherwise employed to affect a business solution.

Some plans may be used where marketing (used here as a generic term) is said to be paramount to the financial success of the business. To this end, more emphasis is placed upon specific marketing strategies employed with a limited or a cursory financial analysis. Whereas, this author would advocate a "*business plan*" planning strategy where the methodology employed considers five important criteria.

1. To use text book "marketing planning" and "marketing strategy" techniques suggested by leading authors but to adapt them accordingly in an iterative manner.

2. To then structure the marketing plan into a format where its synopsis is used for "*business plan*" presentation techniques* but where its expanded format is used to provide corporate strategy guidance associated with business planning, implementation, measurement and control management.

3. To use a financial summary as the point at which any marketing technique employed can be seen to impact upon. Perhaps more importantly, to provide a degree of flexibility for the structure of the financial summary and it's potential to vary in relation to market fluctuations.

4. To structure a *business plan* that reflects its impact across a given market environment and the business as a whole including the activities of all individual company departments.

5. Last but not least is the need to create revenue evaluation techniques using a chronological and iterative process in respect of available or planned competence resource that will maximise our opportunities and limit our risks.

*This suggested methodology may be more attractive to the larger organisation and financial institutions where corporate management may wish to review the projected profit and loss account before analysing how the *business plan* will be achieved.

These proposals will be seen to show how marketing will impact upon business in three specific ways.

- It will allow the corporate management team a simpler method of business strategy identification and its subsequent management for all business types.

- It will create an easier method of generating quality enquiries for small to medium organisations that do not employ any sales persons.

- Last but not least is the impact that put forward has in respect of company resource training needs. What skills levels does the company currently employ? What additional skill levels are needed and in what way will they benefit the business activity?

The following chapters contain a mix of text and models that describe specific methods of undertaking, producing and implementing an effective *business plan.* Each chapter will conclude with an example associated with the suggestions put forward.
Perhaps more importantly are the directional suggestions put forward in respect of corporate management activity. Yes they are responsible for improving business results but in what way must they connect with their management team and vice versa in a way that will convert optimism and pessimism into realism using a more reasoned approach?

Research documentation where appropriate is used to support the *business plan* techniques in a summarised format. This it will be seen is particularly useful in relation to the examples put forward. In that, the example is based upon a fictitious company that employs technology within its market mix. Technology has the potential to create much market turbulence and as such the need to maximise our knowledge base will serve to simplify and limit any potential risks.

Research data has been attached to the end of each separate chapter. This action it is hoped will make it easier to digest the importance of each chapter without having to consider who states what and when it was stated. Unless otherwise stated all that is given within is the work of the author. Should any mistakes have been made apologies are tendered in advance and corrections (if any) should be sent via the publisher.

Contents

Business Plan Strategy Review

To produce and implement a successful ***business plan*** there is a need to identify a planning strategy of some description. This is expressed in a relatively simple format as shown in Fig 1.

Where are you now?
Where do you want to be?
How are you going to get there?
When do you want to get there?
What must you do when you get there?

Fig. 1 Planning Strategy

Normally found in many marketing text books in an array of different formats this statement irrespective of its generic origin requires further definition to make it more effective for any business. It needs to have a relationship between current business activities and those that are intended for the future. This relationship is expressed as:-

- Long-term strategy
- Operational strategy

Here long-term strategy considers where you want to be in the future; say three to five years and its association with figure 1. Operational strategy on the other hand reflects what one must do to meet the needs for the current projected financial year and its association with figure 1. We must also consider any impact upon operational strategy arising out of long term strategic needs.
To this end any changes to operational business activity to support long term strategic needs must not detract from operational trading projections unless there is a rational need to make any changes.

1.1 Where is the company now?
Text book authors suggest that businesses can be assessed in two ways. They are either "Sales Orientated" or they are "Marketing Orientated". Readers can gauge their own business activities using the descriptive features given in Fig 2.

"Sales Orientation"

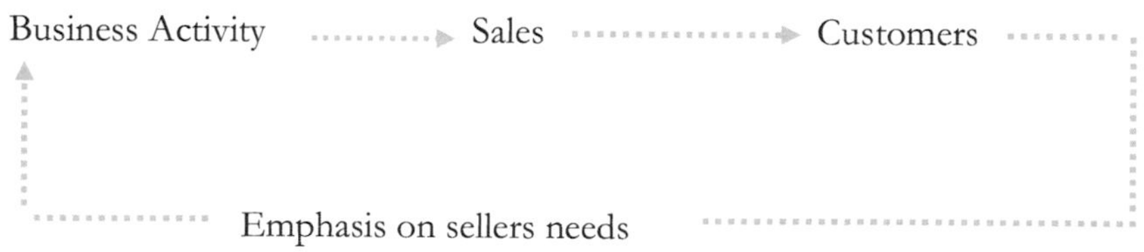

Business Activity → Sales → Customers

Emphasis on sellers needs

"Marketing Orientation"

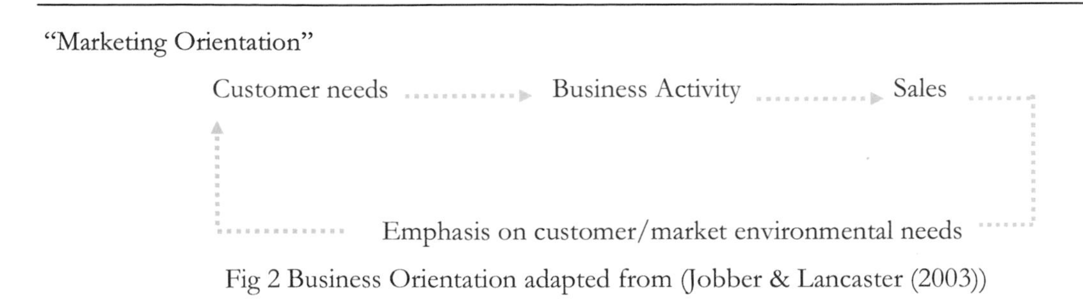

Fig 2 Business Orientation adapted from (Jobber & Lancaster (2003))

So which is the best business strategy to employ and why?

An emphasis on market/environmental needs rather than the seller's needs is seen to be the differentiating factor. Perhaps more importantly this same differentiating factor will be seen to have a dynamic effective on our strategy review. Sales on the other hand will attract a more static business orientation. One that is heavily reliant upon selling techniques and its feedback process to create or formulate business management operating philosophy.
This author would argue that while the sales orientated strategy will produce a return on investment it is a market orientated strategy that would be the best practise to pursue. Here are several reasons why this could be so.

- Selling conjures up many different ways of analogy of its meaning. Marketing on the other hand begins with textbook supported generic meaning and functionality and attracts additional reasoning through specific journals. Anyone can access these journals if so required to support their own specific business activity. Selling is not entirely supported on the same basis and its reference to business criteria nearly always take into consideration *marketing* related philosophy or fact.

- Marketing has evolved and continues to evolve into a broad subject to an extent where substantial data is available for different business sectors. This data, if used correctly has proven effective when comparing companies who operate using marketing techniques in relation to those who don't. Indeed, professional companies are employed to undertake research in this regard the output of which is assessed by textbook authors, they state:

> "Detailed analysis of the research results suggests that the key differentiators between the top performers and the rest are probably not so much issues of strategy or tactics as questions of commitment, cultural consistency and leadership where marketing is concerned. In specific terms, the top performers show a greater marketing grasp and a stronger and more clearly worked-out commitment to marketing principle."

> Adapted from Baker 2000 (P524) Marketing Strategy and Management (3rd ed)

- Most if not all companies operate in a competitive world. How would a "sales" orientated company react to a competitor who is market orientated? Here the orientation alluded to is that where marketing orientated companies look to market needs or wants in relation to their resource base to create their competitive advantage?

- Marketing is very much a cohesive activity within a business environment. People often refer to the "Sales and Marketing" department. Whereas marketing if applied correctly could easily be expressed as the "Marketing, Finance and Production" department or a mix of others.

  In reality, it will be seen that marketing is the function that fuses together the company strategy in respect of company departmental *business plan* needs. Conversely, it also acts as the diffusing activity in respect of disseminating any strategy across the business environment. The "Sales" department would then be responsible for achieving order "in-take" projections. In the same way other departments, for instance have their own individual departmental objectives that must be met.

- The role of the "sales" person is definable in many different ways. It could be as a representative or executive or engineer or commercial traveller or indeed a business development individual. But in what way do they function in respect of business needs and/or market expectations?
  Do customers see these same individuals as "order takers" or someone they can contact when all may not seem as envisaged?
  Are they given a product and told to find markets and to then pursue business prospects?
  If so, are these same sales individuals' mind readers?
  Do they know what prompted the inventor of the product to produce the goods on offer, in particular the value it creates for consumers or purchasers?

  Most probably the business will expect these same individuals by way of their experience or otherwise to conjure up magical ways of achieving or exceeding sales targets. We must also consider the number of sales training courses/aids that are available that promote a varied mix of improved selling techniques.

The art of Selling must incorporate knowledge of the protocol employed for undertaking a meeting of minds between seller and buyer and users for different markets as a basic requirement. Successful selling techniques are those where the seller and buyer and users for different markets undertake a meeting of minds in respect of product or service value creation differentiation relevant to competitive forces. We must also consider delivery, installation and maintenance activities and their respective differentiating factors for certain market segments as an addition to the latter definition.

Clearly the "protocol employed" will vary for each market segment. This is because the market variables and the boundaries that they operate within will not be the same for each market segment. While this is discussed further within the following chapters, it is the sales person by way of the

business contact that they undertake who are more aware of differences in protocol requirements across the segment mix.
More importantly perhaps is the manner in which the more relevant variables may change; changes in the purchasing team or process or exchange rate for instance. It follows that market data derived via a "sales" based function will be that which is current at any given time. Marketing data on the other hand may use information that was produced with a different set of variable definition.

Here the emphasis is all about the methodology associated with our understanding of market data in relation to our approach to marketing. Marketing will allow any changes in variables to be absorbed in a chronological format compared to a high level of business uncertainty and spasmodic change for a "sales" based organisation. All of which we can say will have a significant impact upon how one should define the employment status of a sales person.
Any lack of interactivity diffusion between business and markets all serves to create turbulence within the business processes employed by both the sellers and users. Vendor processes become fragmented, users become frustrated; all of which promote new market entrants and/or an increase in competitor activity and its corresponding impact upon sales and profits.
This turbulence it would seem could emanate from how we orientate our business activity (Fig 2); figure 2A refers.

Cause	Effect on Sales Orientation	Effect on Marketing Orientation
There will be people or organisations who will buy your products as they exist now.	Sell in some of the markets some or all of the time.	Will know and understand what the opportunities are and where they exist and will respond to needs and wants in a manner better than the competitors. It follows that there is:- - Sales - Good Market Position - A high level of credence.
There will be people or organisations who can be persuaded to buy your products as they exist now.	Sell in some of the markets some of the time.	
There will be people or organisations who cannot be persuaded to buy your products as they exist now.	No sales, poor market position and lack of credence.	

Fig 2A – Business Orientation "Cause & Effect"

There is also the impact upon management intervention to secure specific contracts or to assist with frustrated contracts that cannot be resolved by the appropriate responsible person. Does it not follow that if a *business plan* strategy is employed that limits competitive behaviour then the dynamics of our business is more controllable?
Business dynamics is a rarely used term if at all. It is a statement that considers a vibrant (or otherwise) self motivated commercial activity where the foundation of all processes and procedures

follow a distinct pattern of balancing business performance in relation to market trends or technological advancements.
One positive aspect of an awareness of business dynamics is the limitation or removal of micro management activity. That is, personal orientated needs associated with individual or micro activity rather than what is best for the business.
Similarly, it allows us to better understand market needs or wants; or to gauge more effectively technological push techniques owing to a greater knowledge of the individual purchasing or using the goods on offer. All of which cannot be effectively incorporated within our business activity without a modicum of marketing science application.
Clearly, it is possible that certain readers may be budding entrepreneurs eager to start their own business. This being the case such readers may wish to review this particular sub chapter in the form of the potential hazards that could impact upon their future business activities.

1.2 Where do you want to be?

Aspirations exist within most if not all of us. Such a question could imply many things, such as.

- You may be considering new products for new markets.
- You may be considering new products for existing markets.
- You may wish to create new markets for existing products.
- You may wish to enhance the existing product mix for existing markets.
- Undertake Cost reduction exercises.
- Quality Improvements.
- Reliability improvements.
- Increased or more efficient methods of manufacture.

Furthermore in what way can we achieve these aspirations without any radical undertaking? Similarly, if a radical undertaking is suggested then in what way can we limit its impact upon the business? It is a question that should be considered in the same context as “How are you going to get there”. Ergo, in what way should we produce our *business plan* and what must be done to manage its implementation? Moreover, when must it be completed and what do we do when we get there?
A *business plan* if produced in an effective manner will allow the business to identify how it can achieve its visionary aspirations. More importantly, however it will allow the business to offset these same aspirations to balance market demands and competitive forces.

1.3 Chapter Summary

All *business plans* will require a level of approval for the manner in which it has been structured and its intended application. Suffice it to say that *business plan* content must be produced in a concise and simple to read format. Each chapter given within this book will conclude with suggested methods of chapter definition/clarification presentation or descriptive techniques to be employed.
For this chapter there will be a need to identify the strategy review process employed to construct the *business plan*. Suggested documentation should be limited to:

- An opening page giving the name of the company, *business plan* for the year/s, date of submittal and details of the person/s responsible for producing it.

- Copy the headings given in Fig 1. List below each bullet points the factors that have affected the review process. For instance:-

    - Where are you now – No 3 in the market place?
    - Where do you want to be? – No 2 by year? - No 1 by year?
    - How are you going to get there? – Data associated with this question will form the basis of the marketing plan employed.
    - When do you want to get there? - Data associated with this question will be dependant upon the output of the marketing plan employed.
    - What will you do when you get there? – Data associated with this question will be dependant upon the objectives and strategies determined by the marketing plan (discussed further hereof).

Assess and define (using Fig 2 & 2A) how you gauge your business environment. The form of assessment should be undertaken using a "sales based data" and "marketing based data" approach. Both will need to be undertaken in relation to competitor activity and as a minimum should include typical data associated with each main heading, It will not be possible to provide all of the prominent data until one has completed the remaining sections.
For instance:-

"Sales" (Depicted in Pie Chart format or similar)

- Market size by volume and financial terms (projected for the year)
- List by segment
- Available market (Defined by the "marketing" criteria given below)
- Target market analysis (Defined by the "marketing" criteria given below)
- No of competitors
- Competitor share
- Competitor growth or decline (may require details from the previous years plan)

"Marketing" (Depicted in Pie Chart format or similar)

- Market wants/needs (identify No. of segments and their total value)
- Segment analysis (provide total market value for individual market needs)
- Company market mix
- Available market (list by product and growth & decline)
- Target Market (list by product and growth & decline)
- Competitor market mix
- Available competitor market (list by product and growth & decline)

- Market growth

In most cases the "Sales" analysis suggested will already be known (or anticipated) by the business management team. This method of approach it is argued considers more a means to an end. In other words we could contend that this type of business management is more of a "reactive" business approach. One that is static and subject to constant revision. The use of a "marketing" philosophy it will be seen provides for a more reasoned "proactive" and "reactive" business attitude.
It follows that and indeed we can conclude this section by stating that a marketing based review process will provide *business plan* recipients with a more detailed market based study and its subsequent positive impact upon the business projections.

1.4 Chapter Example

As this is the first chapter we must start with a fictitious company. For the purpose of simplification the company is described in Fig 3:-

Company name:	Imperial Ventilation Systems
Premises:	Rented (from local council) Factory unit with office space
Products:	Cooling & Heating Ventilation Systems
Chief Operating Officer:	Eugene White
Management Team:	Hayley Jones - Production Manager
	Manfred Brown - Sales Director
	Jonathan Pryce – Engineer
	Patrick Singh – Accountant
	Production staff of 16 persons
	HR, Legal and Advertising is undertaken by service providers.

The company has been operating for five years. It began when Eugene, an engineer and Manfred had both worked for a large ventilation company. Manfred was forever pestered by the sales channel within which they operated for a mid range product the large ventilation company would not pursue.
Manfred took time to estimate probable sales of the mid range unit and discussed the potential with Eugene. Eugene was equally enthusiastic as he had technological ideas that did not fit in with existing corporate objectives. Moreover, his position within the company had become untenable owing to poor trading conditions at the time. It began as a two man operation but has evolved into the major supplier within the UK of a given mix of products.
Eugene and Manfred as major share holders of the company have become embroiled with the day to day activities of the business and have to a large extent become distant to market activities. Moreover feedback from the sales channel would suggest the advent of competitors within their target markets.

Fig 3 Company example & Structure

Eugene and Manfred have notified the management team (as given above) of their intention to undertake and produce a formal *business plan.* The team has been given four weeks to produce the plan. Hayley has been nominated as the responsible person for gathering and producing all of the information.

Using the suggestions put forward within this chapter we can begin our *business plan* as follows. The front cover of the plan would comprise of a simple heading as shown in Fig 4:-

Fig 4

Confidential Document

Imperial Ventilation Systems (IVS)

Business Plan Proposals

Year: xxxx/xx

Produced by: Hayley Jones – Production Manager

The next page is used to describe the philosophy employed to produce the *business plan.* This is given as Fig 5:

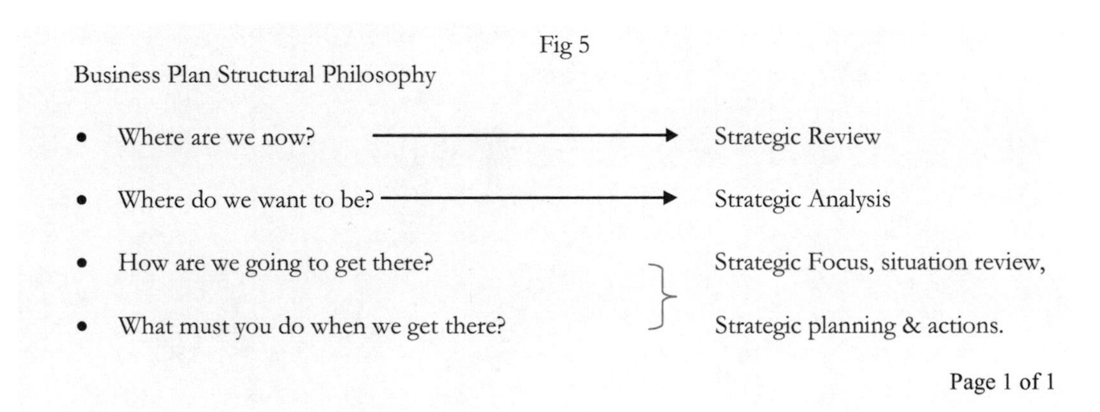

Fig 5

Business Plan Structural Philosophy

- Where are we now? → Strategic Review
- Where do we want to be? → Strategic Analysis
- How are we going to get there?
- What must you do when we get there?

} Strategic Focus, situation review, Strategic planning & actions.

Page 1 of 1

It can be seen that the data associated with Fig 1 has been annotated particularly in relation to the question where do we want to be? Here there is a general statement for the reason for the *business plan.* It follows that some form of analysis is required to achieve the objective put forward. The annotation provided indicates to all that a process of some description must be used to achieve an end to the aims prescribed by Eugene, hence the marketing plan put forward.

We then review each of these bullet points on separate pages; figure 6 overleaf refers.

Fig 6

Strategic Review – Where are we now?

Business Activity

The business philosophy employed at the time of origin had an emphasis on customers needs. Technology usage created market demand and a growth in the product mix. During the last two years the company has evolved on the basis of supply and demand in respect of production availability.

Information source: Major customer survey during April xxxx

Page 1 of 1

For this particular statement it is suggested that all of your major clients be contacted to seek their views in relation to the statement provided. Records should be kept for future reference.

Additional sheets will now be required using pie charts or similar to identify from a statistical viewpoint "Where are we now? The data used was obtained from the business owners who in turn received the information from the distributors sold to and from the allied industries association linked to their industry; figure 7 refers.

Fig 7

Strategic Review – Where are we now?

Market Review

IVS is the leading supplier (by turnover) of clean air ventilation systems within the UK. We are active in all market segments using three specific products. Additional resource is used to provide turn key solutions for certain market segments.
Business has been expanding at a rate of 15% (£1.8M) p.a. which has attracted an additional competitor. Profit (consolidated) has however only increased at a rate of 10% p.a.
Future projections show that sales will increase at the same level however profit will decline if we continue with the existing business activity.

Page 1 of 1

We follow this sheet with a market analysis comprising five separate sheets.

Fig 8

Strategic Review - Where are we now?

MARKET ANALYSIS

Page 1 of 9

We begin with an analysis of the total market and the level of trading activity associated with the company product mix.

Here there is a need to provide data in respect of total sales undertaken for the previous year or anticipated for the current operational trading activity. This example also indicates the market share in real or anticipated terms as a percentage of the total market value.

Fig 8.1

Strategic Review - Where are we now?

Market Analysis – Total Market Review

Target Market	Total Value	Product A	Product B	Product C
Heavy Industrial	£5M	£3M (60%)	£1.5M (30%)	£500K (10%)
Light Industrial	£2M	£1.5M (75%)	£300K (15%)	£200K (10%)
Food & Beverage	£1.5M	£1M (66.6%)	£100K (7%)	£400K (26.4%)
Hotels	£1.5M	£500K (33%)	£200K (13%)	£800K (54%)
Domestic – Urban	£1M	£400K (40%)	£300K (33%)	£300K (30%)
Domestic – Sub Urban	£500K	£150K (30%)	£200K (40%)	£150K (30%)
Total	£11.5M	£6.55M (57%)	£2.6M (23%)	£2.35M (20%)

Page 2 of 9

We then provide three separate sheets that identify how each product is operating within the given market segments.

Sheet 2 – Product A

Fig 8.2

Strategic Review - Where are we now?

Market Analysis – Product A

Target Market	Market Value	IVS Share	Competitor A	Competitor B
Heavy Industrial	£5M	£3M (60%)	£1.5M (30%)	£500K (10%)
Light Industrial	£2M	£1.5M (75%)	£300K (15%)	£200K (10%)
Food & Beverage	£1.5M	£1M (66.6%)	£100K (7%)	£400K (26.4%)
Hotels	£1.5M	£500K (33%)	£200K (13%)	£800K (54%)
Domestic – Urban	£1M	£400K (40%)	£300K (33%)	£300K (30%)
Domestic – Sub Urban	£500K	£150K (30%)	£200K (40%)	£150K (30%)
Total	£11.5M	£6.55M (57%)	£2.6M (23%)	£2.35M (20%)

Page 3 of 9

And for Product B

Fig 8.3

Strategic Review - Where are we now?

Market Analysis – Product B

Target Market	Market Value	IVS Share	Competitor A	Competitor B
Heavy Industrial	£3M	£2M (66%)	Nil	£1M (34%)
Light Industrial	£1M	£400K (40%)	£300K (30%)	£300K (30%)
Food & Beverage	£2M	£600K (30%)	£1M (50%)	£400K (20%)
Hotels	£1M	£200K (20%)	£500K (50%)	£300K (30%)
Domestic – Urban	£2M	£400K (20%)	£300K (15%)	£1.3M (65%)
Domestic – Sub Urban	£500K	£150K (30%)	£100K (20%)	£250K (50%)
Total	£9.5M	£3.75M (40%)	£2.2M (23%)	£3.55M (37%)

Page 4 of 9

And for Product C

Fig 8.4

Strategic Review - Where are we now?

Market Analysis – Product C

Target Market	Market Value	IVS Share	Competitor A	Competitor B
Heavy Industrial	£1M	£500K (50%)	£200K (20%)	£300K (30%)
Light Industrial	£2M	£1.2M (60%)	£300K (15%)	£500K (25%)
Food & Beverage	£1.5M	£1M (66%)	£100K (7%)	£400K (26.4%)
Hotels	£1.5M	£300K (20%)	£800K (54%)	£400K (26%)
Domestic – Urban	£1M	Nil	£600K (60%)	£400K (40%)
Domestic – Sub Urban	£500K	Nil	£200K (40%)	£300K (60%)
Total	£7.5M	£3.0M (40%)	£2.2M (29%)	£2.3M (31%)

Page 5 of 9

Using the statistics derived from the three latter sheets we can state the following:; fig 8.5 refers:

Fig 8.5

Strategic Review - Where are we now?

Market Analysis – Product Mix Vs Market Volume

Total Market = £28.5M

Available Market = £27.0M

Total Sales = £13.3M (49% of available market)

Page 6 of 9

We now need to provide statements in relation to why we have achieved the success depicted within the previous sheets. This is undertaken using a market needs Vs company market mix analysis. This can be a simple or complex statement depending upon the *business plan* author and intended recipient needs.

A separate sheet should be provided for each target market. However, to simplify this exercise only two sample market assessments are provided.

Fig 8.6

Strategic Review - Where are we now?

Segment Analysis – Product A Vs Segment Needs

Heavy Industrial	% of Market Demand	IVS Mix	Competitor A Mix	Competitor B Mix
Price	60%	Yes	Yes	Yes
Quality	65%	Yes	Yes	Yes
Reliability	70%	Yes	Yes	Yes
Turnkey	95%	Yes	Yes	No
After Sales	70%	Yes	Yes	Yes
Delivery	60%	Yes	Yes	Yes
Feature A	90%	Yes	No	Yes
Feature B	40%	Yes	Yes	Yes
Feature C	10%	Yes	Yes	Yes

Page 7 of 9

Fig 8.7

Strategic Review - Where are we now?

Segment Analysis – Product C Vs Segment Needs

Hotels	% of Market Demand	IVS Mix	Competitor A Mix	Competitor B Mix
Price	90%	Yes	Yes	Yes
Quality	65%	Yes	Yes	Yes
Reliability	60%	Yes	Yes	Yes
Turnkey	45%	Yes	Yes	No
After Sales	70%	Yes	Yes	Yes
Delivery	50%	Yes	Yes	Yes
Feature A	20%	Yes	No	Yes
Feature B	10%	Yes	Yes	Yes
Feature C	10%	Yes	Yes	Yes

Page 8 of 9

Last but not least is the need to provide information in relation to market growth and decline as well any static activity.

Strategic Review – Where are we now?

Market Growth Analysis

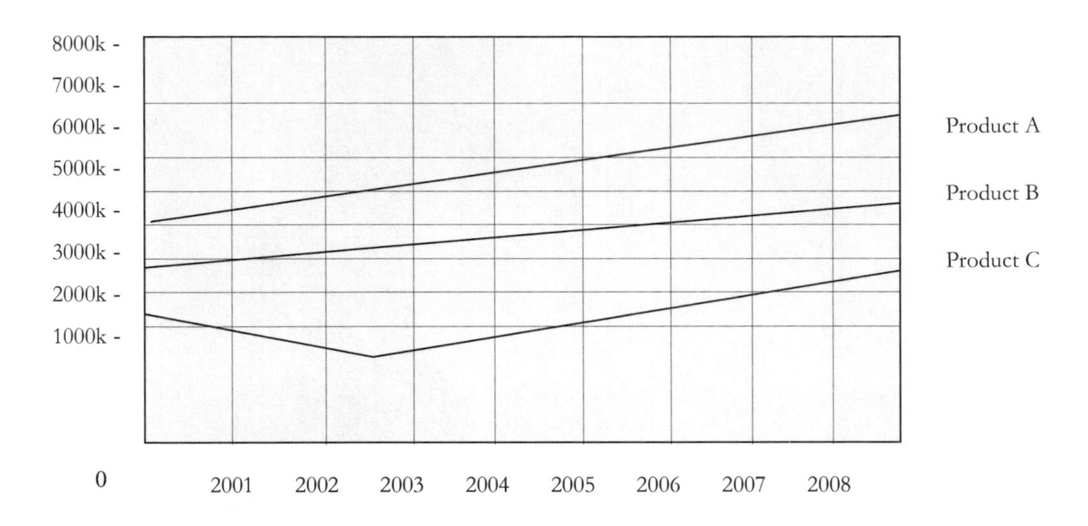

Market Growth – Heavy Industrial Market Segment

Fig 8.8 - Page 9 of 9

Ideally, prudence would suggest that competitor activity anticipated over the forthcoming years should be listed. However, Eugene and Manfred have lost touch with the market and can no longer predict through facts or otherwise what competitor activity represents in terms of anticipated market share. A fact it must be said is apparent for many industries.

The next two sheets are self explanatory as follows:

Fig 9

Strategic Analysis

Where do we want to be?

Page 1 of 2

Fig 9.1

Strategic Analysis – Where do we want to be?

Operational Objectives

To sustain market leadership by reverting back to a business activity based on using market led strategic objective analysis.

Long Term Objectives

Undertake product mix analysis and development in relation to target market segment macro environmental needs and the opportunities arising out of global warming.

Page 2 of 2

The "how are we going to get there?" phase of the *business plan* is used to begin the example provided at the end of section 2 – The *Business Plan* Planning Strategy. The "when do we want to get there?" and "what will we do when we get there?" phases will follow towards the end of the book.

Supporting business plan research criteria

The research criteria put forward serves not only to put forward a synopsis of that we will encounter during the following chapters but will create a reason for adopting the strategy used to create and implement the *business plan.*
Text books and research journals are available in great numbers and an equally vast range of subject matter. Many hours, if not days and months could be expended researching data that may give rise to a host of arguments that has the effect of raising too many uncertainties about one's initial aims. Unless we limit these uncertainties the effectiveness of our *business plan* and its implementation must surely be doomed.

The IVS product mix is very much dependent upon the application of technology. The level of application (irrespective of vendor) will in turn create many variables that will impact upon markets and the macro environment. The variables that exist cannot be easily recognised. Nor can one gauge effectively the boundaries that these variables operate within when reviewed in their text book presentation format.

Technology it could be argued is fundamental in nature. Batteries (or accumulators) provide electrical energy but have different uses depending upon application. They are supplied by different vendors who employ different methods of differentiation that gives rise to a number of variables. Boundaries however will remain the same, that is, the ampere hours rating of the battery is limited by technology and/or its application. But why is such a statement important?

Let us not forget that all managers and supervisors have company personnel that are subservient to them. More than often this same staff will require some form of guidance for them to complete the task asked of them. It could be argued that the quality of the latter is dependant upon how the initial request or order was given and to whom. Put simply, how have we explained the boundaries we wish them to adhere to? Variable definition however will never be fully known as they are subject to change depending upon how an individual views or adapts to them.

This would suggest that any reference support provided by text books hereof should separate known or anticipated boundaries at the outset. The methodology employed to achieve this aim was to observe the authors text in block diagram format to see how these variables interact with other variables in relation to the boundary data associated with them. Each author may use a technological definition to describe a different marketing philosophy or principle, which has the impact of detracting from goal relevance and/or operational validity (Thomas & Tymon (1982)) if not complicating the whole of the subject matter.

Operational validity in terms of the point of reference was achieved first by grouping the author's subject criteria in relation to the technological impact and to then group the common marketing topics where technology is seen to have an impact upon the assertions or otherwise put forward by the authors. The net result of this research is the emergence of twenty-six marketing themes for consideration as shown in Table 1 below.

Clearly, the potential exists to impart most if not all of the data associated with these themes. Indeed, those readers employed within a marketing based organisation may wish to ensure that the *business plan* reflects adequately the needs of each relevant theme.

1	Technological reasons for new textbook text
2	Technological reasons for marketing
3	Technological reasons for marketing & society response criteria
4	Technological reasons that impact upon the creation of customer value
5	Technological reasons that impact upon market demands
6	Technological reasons for market planning
7	Opportunities and Threats
8	Technological impact upon the environment
9	Technological impact upon the marketing environment
10	Technological impact upon consumer buyer behaviour
11	Technological impact upon marketing strategy

12	Technological impact upon marketing strategy resources
13	technological impact upon the marketing mix
14	Technological impact upon "Product"
15	Technological impact upon Portfolio Summary
16	Technological impact upon Competitor Analysis
17	Technological impact upon Competitive Strategies
18	Technological impact upon Market Segmentation
19	Technological impact upon Target market selection criteria
20	Technological impact upon Positioning and branding strategies
21	Technological impact upon Forecasting
22	Technological impact upon Distribution & Sales Policy
23	Technological impact upon Implementation Marketing
24	Technological impact upon Internal marketing
25	Technological impact upon Integrated Marketing
26	Technological impact upon Customer Relationships

Table 1 Marketing Criteria and Technological Relationship

It is not the intention to delve deeply into any of the twenty six criteria given. They are for use as a reference when seeking to expand upon the implementation requirements of the business at hand. Furthermore, the responsibility of adaptation of the criteria rests with the process manager. Executive management will be more concerned with a business over view and as such these same twenty-six themes were reduced to fifteen in terms of their relevance hereof. These are listed in Table 1.1 overleaf together with comments on *business plan* impact associated with them.

Technological Impact	Text Book Author comments
Technological reasons for marketing & society response criteria	It is suggested that marketing does not exist within certain segments - what is the impact upon business plan creation? Kotler et al (2002, Pages.43,56, 68)
Technological reasons that impact upon the creation of customer value	Do the suppliers and the end users undertake value creation analysis if so to what extent? – How does impact upon planning strategy? Kotler

	et al (2002, Pages 21, 34, 369, 393, 394)
Technological Impact upon market demands	In what way does technology impact upon the demands of the market intermediary & end user? – How does this affect the business plan? Piercy (2003, P.59)
Technological reasons for market planning	Is any market planning undertaken by the segment product and service suppliers? If so what procedures will impact upon planning philosophy? Piercy (2003, Pages 74, 81), McDonald (2003, Pages 529, 546, 621), Baker (2000, Pages 79, 86, 137), Hooley et al (2004, Pages 40, 41, 45, 48), Jobber and Lancaster (2003, P.37)
Technological Impact upon opportunities and threats	Are SWOT analyses undertaken? – What is their impact upon a business plan? Kotler et al (2002, Pages 83, 84)) Mcdonald (2003, P.578), Piercy (2003, P.542)
Technological Impact upon the environment	Is it possible to gauge the impact on the environment with the use of current technology? Where do we analyse this requirement within a business plan? Hooley et al (2004, P.105), Kotler et al (2002, Pages 94, 118, 135), Baker (2000, Pages 171, 173, 183, 185, 189))
Technological Impact upon the marketing environment	How does technology impact upon marketing services? Does this impact upon Fig 1 given above, if so how? Baker (2000, Pages 238, 244. 324)
Technological Impact upon consumer buying behaviour	How do the market intermediary and end users react to technology push techniques? Here again does this impact upon Fig 1 above? Kotler et al (2002, Pages 190, 203, 223), Baker (2000, Pages 221, 233),
Technological Impact upon marketing strategy	Are marketing strategies employed? If so does technology have an impact upon the business plan?, Piercy (2003, Pages 289, 308, 347, 393, 437, 457), Davidson (1997, Pages 256, 267), Baker (2000, Pages 56, 59, 64, 81,114, 135, 510, 513), McDonald (2003, Pages 257, 261, 587)
Technological Impact upon marketing strategy – resources	Does technology have an impact on market segment resources? Can we prove that this has a direct impact upon the profit and loss analysis? Baker 2000, P.251), Hooley et al 2004, Pages 140 – 171),
Technological Impact upon "product"	How are current technology stocks applied to product development and what impact does it have on the macro environment? Is the impact upon P&L direct or indirect? Baker (2000, Pages 329, 331, 334, 339, 342, 343, 347, 353, 354, 359). Kotler et al (2002, Pages 497, 501/3, 517, 520, 523), McDonald (2003, Pages 187, 203), Davidson (1997, Pages

	468/9), Hooley et al (2004, Pages 468/9, 470, 472, 475, 480/1),
Technological Impact upon portfolio summary	In what way does technology impact upon the market segment portfolio? How do we manage this as part of our business plan planning strategy? Kotler et al (2002, P.84), Davidson (1997, P256), McDonald (2003, P.583), Hooley et al (2004, Pages 70, 73)
Technological Impact upon market segmentation	Do market segmentation principles have an impact on your market segments? How does it impact upon our analysis of Fig 1? Hooley et al (2004, Pages 337-357)
Technological Impact upon forecasting	Do the market intermediaries and end users provide segment data and is this same data obtained by the suppliers? How do we use any such data as part of our business plan? Hooley et al (2004, Pages 237-253)
Technological Impact upon distribution and sales policy	How does globalisation impact upon distribution and sales? How is it accommodated as part of our business plan? Baker (2000, Pages 397, 402), Piercy (2003, Pages 567, 569), Kotler et al (2002, P.166) Jobber and Lancaster (2003, Pages 91, 173),

Table 1.1

The most notable conclusion that can be derived from this sub chapter is the amount of data that can be collected, collated and stored for primary research for the market audit (chapter 5.0). Technology does have a substantial impact upon marketing but it is not assessed in micro or macro environmental terms by the authors. They write about specific areas of marketing such as "branding" or "marketing strategy" for instance and offer a technology based critical review process to support their own assertions. It is suggested that this allows other researches to pursue a given marketing subject and to then expand if need be upon the authors own assertions with additional research using alternative research journal data. This postulation was considered as the basis for this *business plan* design process.

The *Business Plan* Planning Strategy

2.1 Strategy Formulation

The benefits of implementing a *business plan* planning strategy are somewhat numerous. This will become more apparent within the following chapters. One major benefit is that the finished product will allow the corporate management team to have a better understanding of the key issues that impact upon the business and to measure and manage that impact accordingly. It follows that the difficult subject of corporate strategy and its analysis is made simpler.

It is suggested that the application of any planning strategy should have a level of recognition of the business arena in question. In marketing terms this is expressed as variable recognition and the boundaries that they operate within. Indeed, marketing is the only science available that will allow us to gauge effectively how these two factors are truly important to any business activity as given within Fig 10.

- The ability to define all of the variables that exist within the business macro environment within which the business operates.
- The ability to define the boundary criteria within which the business variables operate.
- The ability to structure the business and subsequently respond to an alignment of company competences in respect of variable and boundary criteria to maximise sales and profit.

- Where are you now?
- Where do you want to be?
- How are you going to get there?
- When do you want to get there?
- What must you do when you get there?

Fig 10 "The Road to Maximising Opportunities and Minimising Threats".

The text associated with Fig 10 can be made more readily understood if we consider more the meaning of the words used. Variables for instance are those factors that impact upon the business that are unpredictable or are erratic or could simply be those that are subject to change on a regular or non regular basis. However, the same meaning could also imply that they are adaptable and may have a level of flexibility. All of which are important when assessing the requirements of chapter 6 – Marketing Audit.

Boundary definitions associated with any variable from a perspective point of view seeks to establish the borders or limits attributable to a variable. For instance, an electric light bulb provides us with light. How much light is dependant upon the wattage of the bulb purchased be it 40watts or 60 watts or whatever wattage sought. In this case, the light emitted is the variable whereas it is the bulb wattage and its method of connectivity that is the boundary definition criteria for that bulb type.

Notwithstanding the latter most if not all businesses will have a specific level of exposure to differing individual personalities that must be borne in mind. Thus the potential exists for boundary definitions to be more complex. To some extent the Marketing Audit chapter seeks to simplify the latter but experience would indicate the need to make assumptions associated with a given boundary as discussed within chapters 7 & 8 – Formulating Strategies and Forecasting & Budgeting respectively.

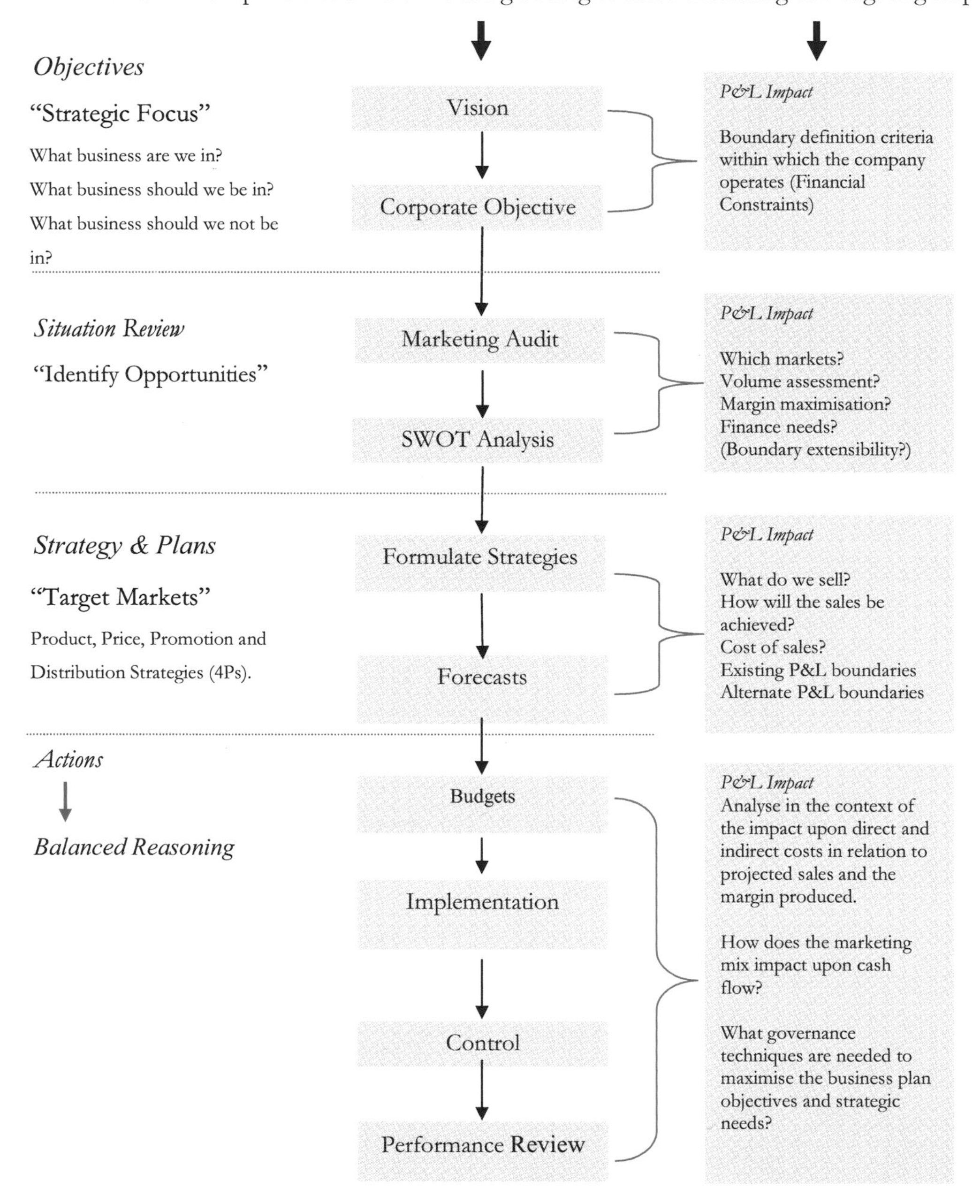

Fig. 11 Suggested *Business Plan* planning strategy adapted from McDonald, M (2003. p.568)

Using that depicted in Fig 10 we can begin to detail a typical *business plan* planning strategy in its basic format. This it can be seen takes the form of a modified marketing plan. Most text book authors suggest that such a plan should consist of four major constituent parts and that these four parts be then broken down into ten separate factors for consideration. For the purpose of this edition the marketing plan as adapted to a *business plan* planning strategy as provided in Fig 11 is suggested.

How each of these ten factors is used within the *business plan* and their impact upon the business is analysed in the following chapters. It is to be noted that the potential exists to revisit certain chapters after reviewing a later chapter and to change ones aspirations depending upon the individual choices made. We refer to this as an iterative process which is not uncommon for successful business operations.

Similarly, it has to be borne in mind that any *business plan* in paper or presentation format must have the ability to be implemented and measurable in a successful manner. Here the management of the marketing plan undertaken by business owners and managers alike is crucial to its success, in particular the implementation of the strategy versus the skills used to implement them. If a sound strategy, matched by good implementation skills exists, then we should expect success in meeting growth targets, profits, and so on, provided that the expectations in setting these targets are realistic.
If a weak strategy exists combined with poor implementation skills not only does the plan risk failure but there is a danger in investing monies in non worthwhile causes.

Clearly any strategy must be seen to fit within the business capabilities, the systems employed and the manner in which the business is structured.

For those of you who may already be marketing graduates, it will be noted that the basic marketing plan has been modified to reflect the impact a given marketing subject has on the P&L. To this end, the application of the plan concentrates more on its individual measurable financial ability rather than academic proliferation. To this end, the book advocates the marketing impact upon business strategy (MIBS) rather than PIMS (Profit impact upon marketing strategy).

2.2 Chapter summary

One will need to assess how long a period of time is available to present the details of this particular chapter. Time permitting it would be advantageous to copy Fig 11. This action would imply to others that all aspects of a *business plan* planning strategy have been incorporated. Thus suggesting not only a professional approach but also that there would be minimal risk associated with any factors that may have been excluded.

An alternative action would be to list the top level activities associated with Fig 11 as shown or similar to that provided by Fig 11A overleaf.

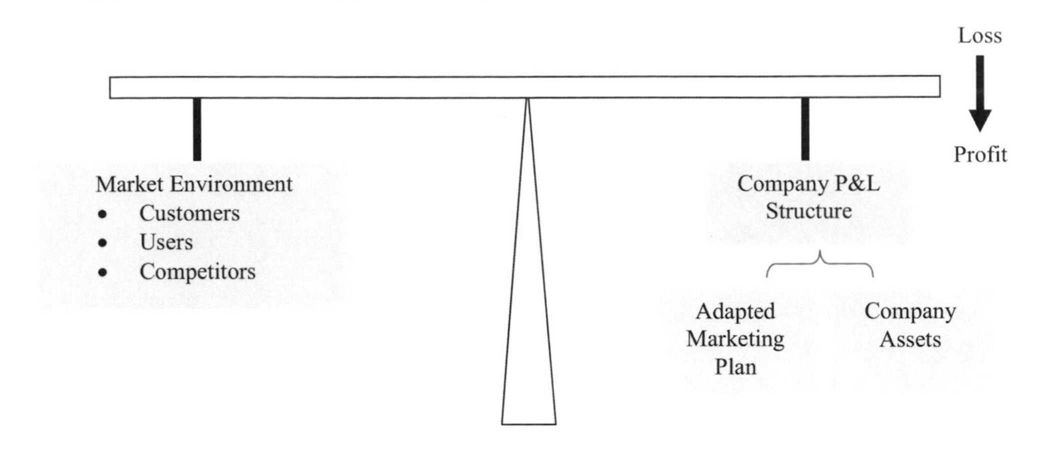

Fig 11A Suggested *business plan* planning strategy (modified)

For those of you having a high level of initiative the latter model can be amended to suit the presentation put forward by way of content or layout or a mix of both. This will become more evident following completion of the work requested within the remaining sections. For instance, you may wish to add to Fig 11A the factors established by the *business plan* strategy that impact upon profit and those that may impact upon loss analysis.

Furthermore, those of you familiar with "Power point" may wish to animate the slide to reflect the factors that affect the balance between the P&L and market demands. These will be determined by the "external factors" and "internal factors" which are discussed further in chapters 5 & 6.

2.3 Typical Example

This section begins where we left off with chapter 1 "namely how are we going to get there?" In other words what process must we implement to meet our needs?
It is the part of the *business plan* where we identify the methodology employed to both select a process and its mode of employment.

Academia is particularly helpful with the first phase of this example. Any analysis (or research activity) of market variable and boundary definition criteria must have a modicum of relevance (Thomas & Tymon (1982)). Our *business plan* submittal Fig 11 simplifies this statement somewhat.

Fig 12

Strategic Analysis – How are we going to get there?

By

"Maximising our Opportunities" and "Minimising our Threats"

Page 1 of 3

Fig 12.1

Strategic Analysis – How are we going to get there?

"How to maximise opportunities and minimise threats"

"By producing a *business plan* that defines the strategic variables associated with our business macro environment; the boundaries that they operate within and to match our competence and resource base with them accordingly".

Page 2 of 3

Fig 12.2

Strategic Analysis – How are we going to get there?

How to maximise opportunities and minimise threats – diagrammatic analysis

Page 3 of 3

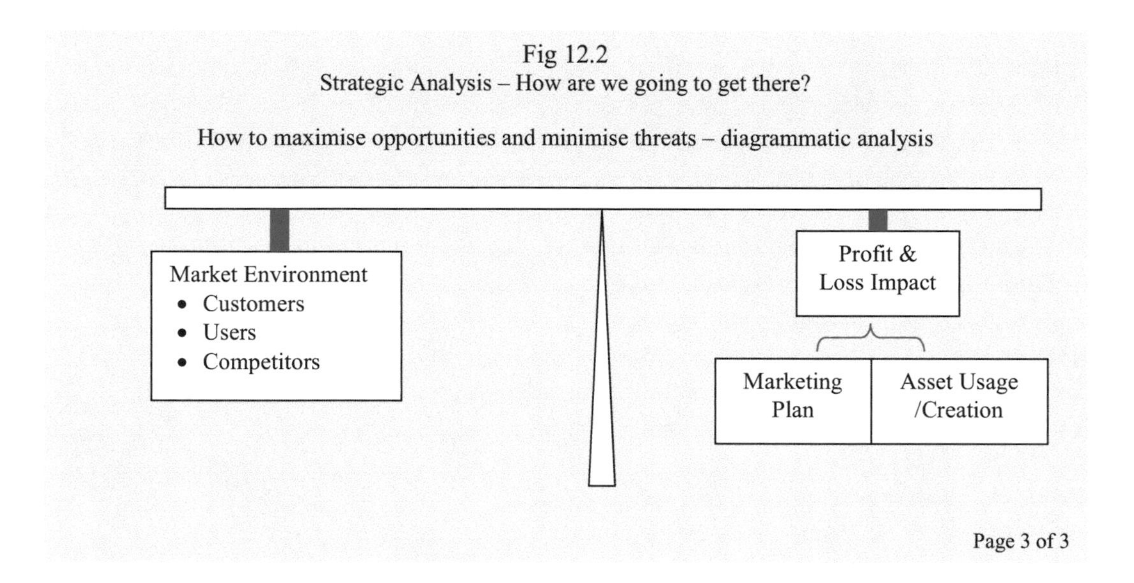

We can now complete this section by introducing the marketing plan as a pre cursor to the remaining sections of the marketing plan planning process, namely:

Fig 13

Strategic Analysis – How are we going to get there?

"The Marketing Plan Planning Process"

Page 1 of 1

Supporting business plan research criteria

Specific contemporaneous research guidelines do not exist in a critically reviewed format. The structure applied to the methodology employed within this chapter follows existing recommendations of research process but specifically adapted to meet the aims and objectives of this book as adapted (Saunders et al (2003, P.7)).
Here one of the main objectives considered was the need to propose knowledge and to test that knowledge using description, prediction and explanation techniques (Meredith et al (1989, P.301)). An attempt was made to observe these three tasks throughout the *business plan* process using predictive, interesting, variable recognition and boundary criterion (Meredith (1992, P.7)). Variables do exists (in great quantities) as do a lack of boundaries for technology based industries. The examples used seek to create a focus for readers to acknowledge their presence and to accommodate them accordingly.

It was observed that criticisms may arise through over or under emphasis of any research associated with the examples provided (Thomas and Tymon (1982, P.345)). The components they suggest for consideration were found to be important particularly within the manner in which the market audit is presented and the example used for its examination.
The examples provided consist of two areas of research; technology and marketing. Differences in the descriptive process for each exist but marketing as a science has progressed that allows the same research material to be used for both subjects.
The objective was to seek what conceptual market plans were available if any that could be adapted for use with marketing and technology. That is any marketing plans that have a technological focus or bias or even with a proven adaptability for technologically oriented companies.

This author accepts that there are two areas of potential difficulty associated with producing *business plans* of this nature where technology has a high degree of focus. Firstly, the association that turbulence has with the understanding of technology in the market place. How can it be removed from any market audit/survey? Secondly, any market survey for technology based markets would be distributed to individuals who are primarily engineers or technology focused individuals.
Here it is argued that readers can establish what it is that motivates individuals. To this end one can recognise that there is a need to create a stimulus that responds to the psychological state of the individuals under review (Mahatoo (1989, P.32)).

Three areas of stimuli can be used; a definition of invention, innovation and new product development but with two connotations (Kotler et al (2002, P.497)). The statement not only acts as a learning scenario but hopefully draws their attention to the fact that the definitions were in fact provided by marketer's (Kotler et al) and not an engineer thus creating an air of importance of marketing. Here one could argue that these three definitions are often not used in the manner in which they are intended.

There is overwhelming evidence to suggest that marketing as a science has failed to produce effective algorithms that permit an assessment of this title for technological based industries using knowledge gained by epistemological methods.
However, sufficient secondary data is available that can be used as building blocks in the creation of a macro focus of a given market segment and the title under review.
The preferred marketing plan under consideration is that proposed by McDonald (2003, P.568). This plan appears to be the only one produced that sets out to identify all of the barriers to be overcome throughout the process in a one page format; albeit, this has been reproduced over several pages as contained in chapter 10.0 hereof.

Slight modifications were made to the plan by way of the impact created by previous research (Erfurt 2002). Note also that "market research" is shown twice within the "Phase 2 – Situation Review" section; this could be a possible misprint by the publishers or author.

It will be seen that this particular marketing plan has the potential to attract too many variables for use with a *business plan* by way of its own entity as well as supporting criteria. It was abandoned in favour of concentrating upon the ten strategic marketing plan contents (McDonald (2003, P.40)) depicted within this chapter. Clearly, the option remains for those readers wishing to pursue the former.

The model definition criteria employ an accepted process (Baker (2000, p. 476) and Jobber and Lancaster (2004, p. 44)). More descriptive text is available but excludes a SWOT analysis (Piercy (2003. p.538)). A different structure is available but the individual marketing principles used are generally the same, it is the method in which they are connected together where the fundamental differences occur (Hooley et al (2004, P.39) and Kotler et al (2002, P.99)).
The chapters provided hereof assess the marketing plan in a hypothetical but well founded format in terms of its ability to meet the needs of a *business plan* proposal.

Vision or Mission Statement

Many businesses will have aspirations for the future. This could be to simply consolidate that which already exists. However even those who have no wish to change will need to let the market know what it is that the company represents other than the product or service that it provides. This part of the *business plan* is used to describe to potential buyers and/or product/service users that which you consider to be your business goal. Several suggestions are provided that could be used as stated or to possibility guide readers into expanding upon that given, these are:-

- To analyse the environment within which the company operates and to examine how the product or service can contribute to a buyers wish to purchase that on offer. This could involve the benefit techniques employed and the net output of them not normally found in competing products or services. For instance, the basic design of a mouse trap has been with us for decades. Similarly, Rolls Royce cars have been with us for even longer periods yet neither actively advertises their goods. The argument being that if a company's marketing is proficient it would follow that a company's mission is already known to the market. (Until such time a higher level of competition enters the market)

- There is a need to define the business if at all possible in terms of the benefits you provide that pleases customers rather than what it is you do just to satisfy a basic requirement.

- There is a need to state any distinctive competences such as the skills or capabilities that the company may possess. Note that it is always better to review the skills in relation to those offered by the competition. Typical values associated with this statement could be levels of luxury or quality or reliability or mix of them all.

- There is a need to analyse future aspirations in terms of what you will be undertaking, what you may undertake and what you would never undertake. For instance Virgin travel transport people by air but does not manufacture aircraft.

- Finally, there exists the potential to review the status of any mission statement for the business that may already exist. Here it is suggested that one carefully reviews the impact upon your customers any mission statement would have. For instance, do any of the services you offer solve problems for certain clients? It could be that reliability arising out of quality is further enhanced by specific methods of service not provided by others. Ergo the products that these certain clients purchase or use may have a longer life span (asset life expectancy?) than competing products.

This data can be further simplified in the form of a descriptive model as shown in Fig 14. The same model can be used to provide something of a difference in meaning between the definition of Vision and Mission. Text book authors suggest that vision relates to future (long-term) strategic intent whereas mission relates to current (operational) intent.

Once completed any analysis of the above should be put into a simple statement that can form a description within the *business plan*. However, prudence will always dictate some form of balance between visionary needs and its impact upon the P&L. Unless of course, if the P&L is produced in a manner that meets with visionary needs.

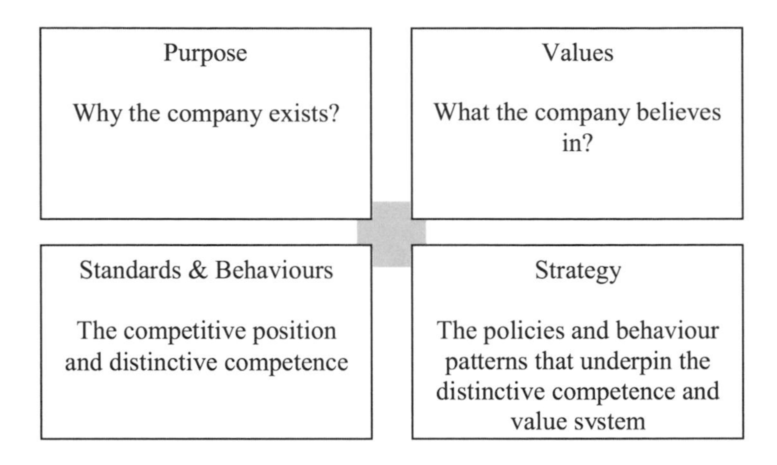

Fig 14 Mission model adapted from Ashridge (Source: Baker 2000)

3.1 Chapter Summary

In the event that no mission statement exists at this time, there will be a need to produce such a statement. Similarly, this book may prompt a redefinition for those businesses already incorporating what they believe to be a suitable statement.

It is suggested that the word "vision" should not be used as operational objectives should take into consideration growth or otherwise to meet long-term strategic requirements. Furthermore, the objective here is the "strategic intent" of the business rather than what you want to be acknowledged for.

Clearly, any such statement may give rise to long drawn out debate in relation to syntax and content. The author would argue that the individual responsible for preparing the *business plan* should first discuss this matter with the corporate management team and/or with his or her colleagues. Additional research supporting material that can be used to provide the company mission statement is given in Table 3.1.

Table 3.1 Mission statement content (adapted from Kotler et al (2002))	
Realistic	Remember that the statement made reflects the company intended mission not any specific product advertising techniques. Be wary of your standing within high growth, high volume or high value based markets. Make sure of your values rather than market position (see also SWOT analysis chapter 7.0). Don't let the

	market drive you to believe that you can improve your leadership or otherwise status. That is do you want to be a market leader for a declining or financially turbulent market segment?
Specific	Ideally, your mission should fit your company (competence and resource base?) rather than any other, in particular competitors. Be specific with the statement, forget generalisations and any contradictions.
Distinctive competences	Re visit your core competence market offering especially the techniques employed for market entry at the outset. How did you differentiate yourself, are the same factors available have they improved, if so why? Always compare your competences with those of the competitors. You may be No. 5 in the market but are known to have a more efficient delivery service than the market leader. Why not ask your key customer base what they believe are your distinctive competences or those they believed you should invest in?
Motivating	You need to provide your management team, sales individuals and most other employees with a set of values that they will feel comfortable with. Imagine one of your sales staff with a potential new key client who is questioning your values in respect of the competition. "I'll have to check with my boss" really does not provide comfort for the client. Similarly, your quality and/or reliability have increased your market share by 20% sounds better than you are market leaders. Note also that some market segments will need more motivation than others. It could be that the buying process will involve a number of individuals and that a number of these individuals may need to sell your values within their own internal approvals process.

Business plans and their creation as well as their implementation must have some form of "Mission" statement not just for its strategic intent but also because of its indirect impact upon *business plan* creation, implementation and the corresponding profit and loss statement.

It will be observed that the remaining chapters will give rise to additions and deletions to a given intent. The adoption of a "Mission" statement will most certainly create a more cohesive directional objectivity to a means to an end for those given the responsibility of producing and contributing to the creation of the *business plan.*

Typical Example

The contents of the marketing plan takes into consideration each individual chapter suggestions in relation to the suggested *business plan* planning strategy depicted as Fig 11 (chapter 2.0) and the strategic marketing model depicted as Fig 51 (chapter 10.0).
The major advantage associated with this method of presentation is the reasons associated with each individual marketing plan activity and their impact upon the *business plan.* Furthermore, this method of approach allows the *business plan* author and its recipients to better understand the importance each activity has in relation to business activity.

This being said we can now add chronologically to the presentation material that provided the conclusion for chapter 2.0, fig 15 refers:-

Fig 15

Strategic Analysis – How are we going to get there?

The Marketing Plan Planning Process"

Section 1 - "VISION"

Page 1 of 2

Fig 15.1

Strategic Analysis – How are we going to get there?

The Marketing Plan Planning Process"

Section 1 - "MISSION STATEMENT"

- What business are we in?
- What business should we be in?
- What business should we not be in?

} (Boundary definition criteria? – Financial Constraints?)

Page 2 of 2

Clearly, the three questions indicated above cannot be answered alone by the person preparing the *business plan.* Similarly, there is a need to ensure that the questions are answered and presented in a simplified format.

The example provided in Table 3.2 first reviews the data gathered during discussions with the management team and production supervisors. The table given may comprise of confidential information and as such care must be taken in respect of who the data is presented to.

Purpose	Why the company exists?	The ability to understand the scope of delivered solutions better than the competition.
Values	What the company believes in?	To adapt technology to market needs not force the market to use the technology available.
Standards & Behaviours	The competitive position and distinctive competence.	The ability to adapt to market needs using reliable products of a high quality at a competitive price.
Strategy	The policies and behaviour patterns that underpin the distinctive competence of the value system.	To monitor incremental advances in technology and to adapt them only when they create value or are needed to create competitive advantage.

Table 3.2

Clearly this data can give rise to much discussion. For the purpose of simplification the IVS company mission statement is put forward in a simplified format as:-

Fig 15.2

Strategic Analysis – How are we going to get there?

The Marketing Plan Planning Process"

"MISSION STATEMENT"

"Balancing technology demands and diffusion for environmental needs"

Page 3 of 3

This statement is particularly suitable for internal and external needs. In that it informs the market that the company is customer focused in the use of technology and that their product mix is adapted for specific target market needs.
From an internal point of view it informs all company employees of the type of business they should be pursuing. That is to seek, segment and develop markets where the company competences are easily matched. A major benefit this comment provides is the number of quality sales enquiries such a statement would generate.

Clearly companies operating with a high visionary management team may wish to change their mission statement on a yearly basis.

Supporting business plan research criteria

McDonald (2003. p.41) uses three general descriptions to categorise how the mission or vision statement can be addressed. One relates to what he calls "Motherhood" that is usually found within a given companies annual report and is used to "stroke" the shareholders.
The second category is classed as the "real thing" and relates to the company organisation. Its impact would apply to the behavioural patterns of executives at all levels.

The latter level takes into consideration statements that would apply to a given business unit, which he classifies as a "purpose" statement. Whereas we are concerned with the visionary aspects of the *business plan* in that it is the *where are we now* question that must be answered hence the adaptation of text book criteria for this chapter and the preceding chapters.

Jobber and Lancaster (2003, p. 36) also consider the visionary statement but in a manner that also supports the use of marketing plan depicted in chapter 2.0. Reference is made to senior management asking the question "what business are we in". Moreover, their assessment of this marketing requirement reviews additional criteria not considered by McDonald (2003). To this end, one should consider the status imparted by the statement. One example given considers the microcomputer business where the business could be described as "rapid problem solving".
Piercy (2003, p. 445) on the other hand provides us with what he calls the "Piercy instant mission kit" as well as substantially more data the most moving of which is his model of mission. Although one of the more definitive paradigms it is considered too complex for this application. We must also bear in mind external public relations and the motivation of employees Baker (2003. p.79).

Corporate Objective

This particular chapter has the potential to be the most difficult to assess when preparing any *business plan*. To all intents and purposes any guidance we would wish to seek from text book authors, who themselves may provide analogous data, is somewhat fragmented, difficult to find (within the overall text) and varies between the authors under consideration. Furthermore, one would suggest that the text provided is undertaken using an assumption that the marketing principles within the respective book are employed to their full extent. In reality, the latter will largely be adapted (if used at all but given hereof) and as such academic based corporate objective principles and their usage will vary widely.

One argument that does exist is that corporate objectives and mission statements are closely linked (Hooley et al (2004) & Davidson (1997)). Davidson goes on to suggest that corporate objectives are "more specific and mundane" than mission assessment. These facts notwithstanding the basis for our fundamental understanding associated with corporate objectives can be summarised as the "Financial Constraints" that impact upon the business. These are the constraints imposed by the corporate objectives.

Money, it has been suggested is the root of all evil. In this context the author would suggest it is the root of any chaos or turbulence that may arise from our *business plan* research process. It is a brave person who takes it upon himself or herself to tell the business owner or appropriate senior manager how they should set the objectives for the business. Most probably a reason why limitations in research supported corporate objective data exists? What we can do is suggest to these same individuals what it is they should be considering and on what basis the techniques employed hereof will help them in setting their objectives.

Financial constraints will impact upon the direction a business must follow or adapt to. They must be reviewed in both operational and long term strategic format. Albeit, it must be remembered that operational financial constraints may be structured in a manner such that it will provide for long-term strategy implementation. Thus we can begin to assess our process for identifying corporate objectives.

4.1 Objectives Analysis and Fundamental Concerns

Business owners and/or the corporate management team will already have goals that they will want employees to pursue. Our initial objective here will be to ensure that the management team are aware of all of the factors that will have given rise to personal preference as well as those that may impact upon personal preference.

Clearly, there exists the potential to expand the corporate objective by way of the boundaries associated with each required goal. Ergo a more definitive approach that may apply to the financial constraints associated with any individual and overall goals. This action to a larger extent will begin the process of ensuring all employees understand the objectives set but perhaps more importantly the objectives meet with their concurrence.

Firstly, we will consider the nomenclature we all find ourselves using to describe our business activities. To a larger extent an analysis of the appropriate nomenclature will, as is shown in the bullet

points given below allow us to understand the management process expected when asserting ourselves using the nomenclature explanation provided.
For the purpose of simplification the research provided is copied in its entirety (Baker 2000). Its usage is more as a directional or support aid at this juncture but will be seen to have a high level of *business plan* performance review and evaluation as discussed in chapter 10.0 hereof.

- "A *strategy* is the pattern or plan that integrates an organisation's major goals, policies and action sequences into a cohesive whole. A well formulated strategy helps marshall and allocate an organisation's resources into a unique and viable posture based upon its relative internal competencies and shortcomings, anticipated changes in the environment, and contingent moves by intelligent opponents.

- *Goals (or objectives)* state what is to be achieved and when results are to be accomplished but they do not state how the results are to be achieved. All organisation's have multiple goals existing in complex hierarchy from "value objectives", which express the broad value premises toward which the company is to strive, through "organisational objectives", which establish the intended nature of the enterprise and the directions in which it should move, to a series of less permanent goals which define targets for each organisational unit, its sub units, and finally all major programme activities within each sub unit. Major goals – those which affect the entity's overall direction and viability – are strategic goals.

- *Policies* are rules or guidelines that express the limits within which action should occur. These rules often take the form of contingent decisions for resolving conflicts among specific objectives. For example: "Don't use nuclear weapons in war unless American cities suffer nuclear attack first" or "Don't exceed three months inventory in any item without corporate approval". Like the objectives they support, policies in a hierarchy throughout the organisation. Major policies – those that guide the entity's overall direction and posture or determine its viability – are called strategic policies.

- *Programs* specify the step-by-step sequence of actions necessary to achieve major objectives. They express how objectives will be achieved within the limits set by the policy. They insure that resources are committed to achieve goals, and they provide the dynamic track against which progress can be measured. Those major programs that determine the entity's overall thrust and viability are called strategic programs.
  Strategic decisions are those that determine the overall direction of an enterprise and its ultimate viability in light of the predictable, the unpredictable, and the allowable changes that may occur in its most important environments."

Source: D.R. Melville *"Top management's role in strategic planning"* in Roger A. Kerin and Robert A.Peterson (eds) *Perspective on Strategic Marketing Management.*

These bullet points serve to establish a common acceptable researched approach for a company to pursue.

4.2 Objectives Analysis Factors

Donaldson states that corporate objectives are all about "turning vision into measurable goals". Piercy (2003) as part of his mission statement analysis suggests that there must be a level of strategic direction.

Examples are given as:-

- A need for fast penetration within a given market
- Will accept investment for long term development
- Wants to maintain the current position
- Turnaround the business to create a new market position
- Harvest what you can in margins for minimum cost
- Prepare for divestment

What should be obvious from these examples is the fact that the objectives put forward do not go into great detail. They do not define the limits that the corporate team is prepared to identify, particularly in the form of financial constraints. To some extent, it is these factors that would suggest that we assimilate the corporate strategy with the vision or mission statement. To this end, "what the company will not do".

However, as with most business environments we must identify and understand what factors have created the objective focus. It is possible that some if not all of these factors will allow those responsible for implementing the objectives to have a better understanding of best guess requirements as and when the need arises. It is possible that there are fundamental changes occurring within the market environment. Those that are occurring out of new technologies in a tertiary or a tangential environment, competitor activity not previously acknowledged or any other factor considered crucial to the objective analysis phase of the *business plan.*

We must also bear in mind our previous questions in relation to the prominence associated with the definition of "Sales" or "Marketing" based organisations.

It is the knowledge base used to determine the corporate objectives that will determine the ability to first arrive at goal definition and their methods of implementation. It follows that to a large extent that the remainder of the marketing plan suggested hereof be completed to maximise the thought process employed to arrive at an acceptable corporate objective.

4.3 Chapter summary

This part of the planning process can be summarised as being the "Goals" for the business in question derived by personal choice (selling process) or by a wish to create and provide a business that responds to market needs (marketing process) in a manner better than others.

It is expected that some if not most senior managers will look upon this chapter in awe. To this end, they are either departmental heads or they are more senior who spend their working lives measuring and managing the performance of others. Yet, the data provided must alert even these individuals to the fact that it is the corporate or senior management team who are responsible for providing the appropriate objectives. Any lack of response in this regard will be considered as management myopia that will give rise to business myopia the result of which must be a turbulent business activity.

For the purpose of *business plan* presentation techniques it is suggested that the corporate objectives as they stand at present be submitted but that questions be asked and listed as the basis used to arrive at the current objectives. If the marketing plan suggested hereof identifies other factors that may impact upon the corporate objective they should then be listed in the form of what they are, that is "Factors for consideration".

Stating that the market size is increasing may not cause a change in corporate objective. What factors have caused such an increase? How could it benefit the company and what must the company do to maximise the benefit?

What the objective must foster is a management force that knows their financial constraints associated with pursuing a balanced company reaction to market actions. Typical factors for consideration are given in bullet point format below. This would suggest that any revised corporate objective data needs derived from the marketing plan be "realistic" and "implementable".

- What products or services to offer, which markets to target and how to outperform competitors in those market places?
- There is a need to create a balance between a firm's need for profit and customers need for value.
- The outcome of your *business plan* will impact upon the product/s that you will provide and the customers that use them. However, do not forget that there is a similar potential for products or services provided by your business to impact upon the market.
- Do you want to be "market leaders" and the risks it may encounter and make other similar businesses "market followers"?

It will be noted that it is possible that businesses may wish to revisit the company "Mission" statement following an analysis of this chapter and to amend accordingly.

Typical Example
Hayley has used the same forum associated with "Mission Statement" to analyse and to then agree with Eugene and Manfred the corporate objectives for the company. Eugene has already put forward his objective views in that he wants the company to retain its number one position in the market but to also examine the impact global warming may have on the company product and service mix.

However, we have also established that both Eugene and Manfred have become embroiled with the day to day running of the business to the extent that neither had realised the impact the competition has had on the market and on the business. This was established within the presentation material put forward within the "where are we now?" chapter, namely yes we are market leaders but only by turnover. There is a high degree of competitor activity as is shown with Table 4.1 given below.

Product	Market Segment	Status
A	Hotels	No. 2 after Competitor B
A	Domestic Sub Urban	Joint No. 2 after Competitor A
B	Hotels	Position No. 3
B	Domestic Urban	No. 2 after Competitor B
B	Domestic Sub Urban	No. 2 after Competitor A
C	Hotels	No. 2 after Competitor A
C	Domestic Urban & Sub Urban	No Sales

Table 4.1

The information provided for this example would suggest that the management team review its objectives. The outcome of which is used to create the corporate objectives for the *business plan*; fig 16 refers:-

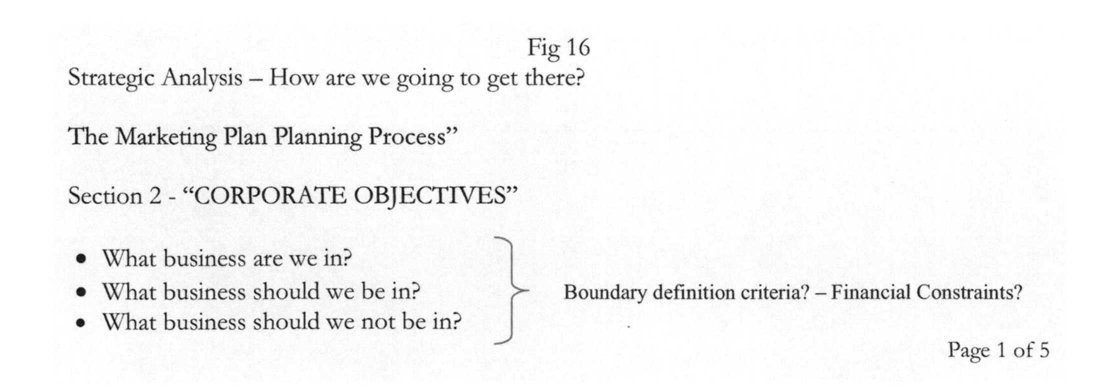
Fig 16
Strategic Analysis – How are we going to get there?

The Marketing Plan Planning Process"

Section 2 - "CORPORATE OBJECTIVES"

- What business are we in?
- What business should we be in?
- What business should we not be in?

Boundary definition criteria? – Financial Constraints?

Page 1 of 5

Fig 16.1
Strategic Analysis – How are we going to get there?

The Marketing Plan Planning Process"

CORPORATE OBJECTIVES – Operational & Long Term Strategies

"To let our markets and competitors guide our technological development and its growth/deletion across the macro environment"

Page 2 of 5

Fig 16.2
Strategic Analysis – How are we going to get there?

The Marketing Plan Planning Process"

CORPORATE OBJECTIVES – Operational & Long Term Goals

"To retain our market status by balancing our product/service mix with long term market and environmental needs and competitor activity"

Page 3 of 5

Note that the latter sheet informs the company employees as well as the market that the company is prepared for any down turn in market trends. Trends it must be said that could impact upon the financial viability of a company.

Fig 16.3
Strategic Analysis – How are we going to get there?

The Marketing Plan Planning Process"

CORPORATE OBJECTIVES – Operational & Long Term Policies

"To create a level of empowerment within the business that will simplify and maximise the manifestation of corporate strategies and objectives"

Page 4 of 5

Fig 16.4
Strategic Analysis – How are we going to get there?

The Marketing Plan Planning Process"

CORPORATE OBJECTIVES – Operational & Long Term Programmes

"The actions and programmes arising out of the implementation of the corporate objectives are contained within the remaining sections of the *business plan*"

Page 5 of 5

Supporting business plan research criteria

No additional research criterion is provided as the chapter text (albeit limited) is self explanatory. What should be noted however is the context within corporate strategy that a marketing plan in respect of a *business plan* has. By using the former to structure the latter it will become evident how any financial constraint will impact upon a company's profit and loss account and corresponding cash flow.

Conversely, it will become more apparent how we can structure our financial support to create a balanced profit and loss statement in respect of a balanced desired revenue assessment.
Indeed marketing as a science and the formidable impact it has upon company structure, operations and performance may not be apparent to many company executives. Its consequent and congruent impact upon corporate objective may not be recognised hence the need to assess and balance the needs of this chapter with our overall *business plan* objectives and assertions.

Marketing Audit

What is a marketing audit? Put simply it is a means of controlling the energy expended to manage the marketing effort. It provides a basis of understanding for business owners to appreciate the environment within which they operate. This in turn will allow these same owners to better understand and identify their strengths and weaknesses in relation to any opportunities and threats that impact upon the business sector. Ergo, it also allows business owners to position themselves within their particular geographical area or market segment by providing products and services that differ and/or are better than others.

The basis for any audit can vary depending upon the text book under review. For the purpose of this *business plan* we will consider that given in Fig 5.1 – 5.4. These particular models allow us to expand upon the subjects listed. It is to be noted that this part of the *business plan* is an extremely important aspect of the plan and as such attracts what could be a complex process.
To this end it involves the use of a whole new set of principles, techniques and substantial personal skills covering a wide range of business and selling tasks. Here we will begin to understand, acknowledge and the need to group the variables and boundary definition criteria previously alluded to.

Perhaps more importantly however is the paradoxical link the audit has in respect of any product or service selling techniques that need to be employed to deliver the *business plan* proposals. The individuals responsible for developing and producing the product or service offering must be made aware of prospective markets and usage criteria to maximise opportunities and to minimise risk. Here the paradox referred to is where "sales" based organisations undertake minimal or no market audit but are reliant upon sales individuals to eke out potential markets and users.
To simplify the process employed four separate self explanatory tables are provided (adapted from Piercy 2003).

Specific marketing terms are used within the audit. In the event that these are unknown to readers, an explanation of the terms is given at the end of this chapter. Similarly, chapter content is considered too extensive to list in its entirety within any presentation techniques employed. Rather, individuals should seek to list those items having a strategic impact in simple bullet statements.

5.1 The market audit (appraising the market)

Focus	Analyse	Objective
Customer needs and buying factors	Customer priorities in the needs to be met through purchase. Customer group differences.	Emphasize customer needs not products, and the difference between potential group differences. Think of every variable that potentially exist other than technical features or otherwise.
Products and customers	Group products by their	Create products and customer definitions

	common need satisfaction characteristics and customers or markets by common characteristics.	reflecting the market place not just your business activity or the internal operations. For example what would your potential customers be looking for in your product/service offering?
Key products	Identify the key products for each customer group/market, the ones you *have* to have to be a player in the market.	Establish customer group/market differences in product priorities. Make a list of the different groups that would be attracted to your business.
Marketing priorities and critical success factors	Evaluation of the most important marketing mix element for each customer group/market and the things the "winners" get right.	Establish relative effectiveness of marketing mix variables and competitive requirements.
Market segmentation	Product groups and customer types and matching criteria.	Define customer-related market segments reflecting differences in customer needs. Look out for any segment overlap.
Company priorities	Compare each match of product and customer to: - potential competitiveness you could achieve. - attractiveness of this business to you.	Isolate areas of low and high priority and any potential niche gaps and opportunities in segments, and match your goals and capabilities to them.
Market sizing and shares	Use static and emerging segments to value market and its trend, and shares taken by competitors.	Place values on segments that will move towards your targets. Remember that variables can be adaptable.
Life cycle and competitive position	Life cycle stage and competitor position in each segment.	Prioritize market segments and niches. Note that all products have a life cycle that will eventually outlive its usefulness. Here again revue any trend data that may be available in particular life cycle trends compared to your competitors.

Competitors	Evaluate: - direct and indirect competition: - enterers and leavers: - major competitors characteristics.	Identify competitive prospects and your shortfalls. Consider the importance of any competitive offerings and the type of businesses offering them across the market segments.
Market environment	Evaluate likely impact of brand changes in markets, law institutions, government regulation, technology etc.	Put planning into broader content of strategic change in the outside world. For instance how effective is the internet when reviewing any potential increase in sales?
Market summary	Across the segments, analyse life cycle stage, value, current and required business direction, priority products and marketing mix requirements.	Collate market and competitive positioning data. This is summarised as follows:- 1. Market Segmentation. • Identify your reasons associated with the markets you have segmented. • Develop profiles for each segment. Remember that buyers have unique needs and wants. See table 5A below for typical consumer variables. 2. Market Targeting • Identify levels of segment attractiveness. • Select the most attractive segments. 3. Market Positioning • How will you position your business for each segment? Develop marketing mix for each target segment.
Market Priorities	Across the segments, analyse your market share and sales projections, chances of	Choose priority market targets. This can be supported by any data provided by any statistical bureau as well as any private

	success and priorities.	research. Revert back to the "market summary" box if so required.
Critical success factors	What *must* be right to achieve customer priorities?	Provide a specific action list in priority format.
Marketing objectives	In priority segments what are the key marketing objectives and how do they relate to sales and the margin they generate and market share.	Isolate major marketing goals, compared to corporate objectives. May require a review of the corporate strategy put forward for the business.*

Table 5.1

* Denotes that sales versus the margin generated in the context of the cost of overheads to provide them will need to be supported adequately by the strategic marketing strategies as further explained within chapter 7.0

Table 5.1A - Segmentation variables for consumer markets

It is to be noted that most businesses will be heavily reliant on specific data to complete this table. Certain business activities will have access to national statistics or those that may be available from a private source. In the event that such data is not available individuals are asked to consider text book alternatives. In that, one can base their data on CUGs (Current Useful Generalisations (Baker 2000) but that any CUGs used be identified as part of the plan. Suffice it to say, the information provided below will at the very least arm the individual with knowledge of the questions that should be asked and from whom in terms of the data that is needed for the *business plan.*
Text book examination reviews this part of the analysis phase in terms of simply auditing the market. However, the potential exists for expanding upon any questions asked to evaluate the standing within a company the individual or individuals has in relation to their ability to implement any change contained within their response criteria. While this requirement may not have an importance for fast moving consumer goods markets it will most certainly impact upon those undertaking audits in business to business market environments.

Variable definition	General Description
Geographic	
Region	Potential customers exist within all regions of the UK as well as globally. Here it is suggested at the outset that one concentrates on any statistical data that may be available or by using CUGs. Available manufacturing capacity and support personnel (competences) will be the limiting factors on the choice of number of regions for consideration.

Region size	For the purpose of *business plan* completion as well as any prospective long term strategy requirements it is suggested that for each region one assesses they be classified in terms of size (volume).
Region Density	Certain regions may have a higher density of prospective customers; could be urban or sub urban depending upon the region under evaluation This information must be borne in mind in relation to the variables put forward hereof and the advertising techniques employed.
Climate	This variable considers the climate of the business environment in question as well as the climate of the regions under scrutiny and their impact upon the strategy employed.
Demographic	
Age	What age group are you targeting?
Gender	Male or Female or both.
Family size	An important variable for specific types of industries in terms of any differentiating factors you may be able to provide. For instance, the density of specific family sizes within a given region may impact upon any point of sale disposition consideration.
Income group	Clearly, the higher the cost of the product and/or services will limit the target market selection criteria. It may also impact upon after sales product pricing and strategy.
Occupation	Clearly an important variable if you produce a product that takes into consideration occupational needs. Here there is a need to consider primary and secondary needs as well as any tertiary activities. An example could be the varied mix of locking systems available as door locks, the types of keys employed as well as the varied mix of key fobs or remote monitoring systems.
Authority	Engineer or manager or neither in relation to wants or needs and their ability to influence or approve any changes arising out the audit undertaken.
Education	Assess if you need to seek data on the educational status of the potential customers that may wish to purchase your goods and what their wants and needs are? Clearly an important factor for those who operate on an international basis.
Religion	Always an important variable particularly with the mix of nationalities

	that may now exist within a given region. Again, possibly a long term strategy area of focus such as prayer rooms in a hotel establishment for instance.
Nationality	Again, another important variable owing to the mix of languages the business may need to support.
Psychographic	
Social class	Working class, middle class or upper class variable data will impact upon many aspects of the marketing mix such as market sizes, volumes, level of expenditure, cost of product, repeat purchase etc.
Lifestyle	Look at how consumer changing needs at different stage of their life can be accommodated.
Personality	Consider product or service benefits that targets consumer personalities. Brand limitations are a classic example that attracts more reasoning with target market selection and definition criteria.
Behavioural	
Frequency of purchase	Here we must consider "market pull" versus "market push" techniques. Does the market require your product? Do you need to alert the market as to why they need to purchase your product?
Benefits customers seek?	Quality, Service, Flexibility, Location or Price or a combination of all?
User status and user rate	Information will be required on who to increase business with and who to secure business from. That is, which segment are regular buyers or users and those segment who do not purchase at all or rarely? Here one must not forget bulk purchase techniques that occur with long intervening periods.
Loyalty status	State in terms of none, medium, strong, absolute and why?
Potential Customers	Assess who to access particularly those who are unaware (including "User status and user rate" variable) of your accreditation as well as those who are aware and informed.
Consumer attitudes	Probably not an easy statistic to obtain; you may need to undertake private research associated with those who are enthusiastic, positive, indifferent, negative or hostile towards your business and product offering. For what reason do individuals have such attitudes?

Table 5.1A; Adapted from: Kotler et al (2002) - "Market segmentation variables for consumer markets".

5.2 The Product Audit (appraising the services you offer)

This model is for use when focusing specifically upon product issues and represents a simpler version than given above. It is particularly important when establishing or determining product positioning, meeting customer needs and product dimensions as perceived by the customer. It will most certainly act as a suitable "*aid-memoir*" by technological staff operating in an environment that is either secretive or detracted from the day to day running of the business.

Prudence would suggest that any audit should be assessed bearing in mind the needs associated with the market communications audit (sub chapter 5.4).

Focus	Analyse	Objective
Competitive performance	In each segment, how well do your products meet customer needs compared to competition?	Identify gaps in matching your products to high priority customers.
Product dimensions	For the critical products or customer needs, how well do you perform compared to competitors on the most important dimensions?	Concentrate on strategic differentiating characteristics of products not just the generic product (basic product design)
Product lines	For the critical products or customer needs, where do you stand against market standards and where are your specific product gaps and deficiencies?	Develop lists of shortfalls and actions to remedy. Also consider comparing this same information with any standards that apply within your industry. It is possible that you may exceed basic requirements.

Table 5.2

5.3 The Pricing Audit

Here the critical issues as perceived by the customer are put forward for consideration.

Focus	Analyse	Objective
Product pricing	Within each segment compare your product prices to key competitors, and your price positioning in the segment.	Identify product price positioning. Establish or create a simple chart where your price positioning elements can be observed as and when the need arises in a simple fashion.
Market pricing	Within each segment compare	Identify market price positioning and

	product prices to key competitors, across the segments and the total market.	relationship with market share. You may wish to seek data from any available national or industry statistics.
Price trends	Examine product price index for past and expected future.	Identify risers and fallers. You may wish to seek data from available national or industry statistics.
Value	Compare perceived quality, price position and market share for your company and key competitors.	Break the "low-price = high sales" perception and look for positioning anomalies.
Price levels	For key products, compare your discount structures with key competitors.	Compare your strategy with competitors and track implication for market share.

Table 5.3

There are "Critical issues" that may apply to product/service providers that may not be readily understood by the customer. One specific critical issue is the variance associated with monetary exchange rates and its impact upon those who operate within international markets.
Variations in exchange rate will affect the product price in relation to international competition prices. That is if you sold a product six months ago and you expect to sell the same product at this time in opposition to international competition who are quoting on the same basis, it does not follow that you will have the same level of price competitiveness.

For instance, you may be a UK based company that markets its products in pounds sterling. You may be selling into the USA and your six months previous price was based upon a rate of £1.00 = $1.60. Whereas the current rate is £1.00 = $2.00 thus the exchange rate has effectively increased your price by 25%. Conversely, your major competitor is of mainland Europe origin that uses Euros for the basis of trading. At the same time the Euro may have devalued thus creating a more competitive price without undertaking any product or service design change. Clearly, such a scenario applies to all currencies.

While such issues may relate more to business operational issues they must be borne in mind at this stage of the plan.

5.4 The Market Communications Audit

Key issues that impact upon this audit are definitions associated with product or service purchasers and users. What sales channels are employed and in what way must marketing be used to allow the diffusion of a companies aims throughout them?

Focus	Analyse	Objective
Brand/corporate	Identify customer perceptions of	Identify the broad communication tasks.

positioning	your company and of your key competitors.	Structure a customer "Feedback" form would be a simple method of gaining data.
Decision-making units (DMU)	Within each market segment, model the DMU, identify the different roles played by the different people, and the relevant messages and media.	Not readily applicable to consumer/commodity based markets unless government regulated. The latter together with the general industrial sector will need to assess the decision making process and the individuals associated with it. Note that separate information can be provided by the author in this regard. See also Fig 40 page 221 "marketing services"
External influencers	Within each market segment, identify major influence sources and your standing compared to the competitors.	Isolate influencer targets for communications and message required.
Media	Within each segment, identify available media of communication and compare.	Analyse and if necessary broaden your view of communications media.
Media performance	Compare your effectiveness in using each medium and expenditure with key competitors.	Relate effort to market share and areas for development.

Table 5.4

Key account management is arguably more associated with this method of market audit and sales management techniques. What should be noted however is the need to analyse the data in the form of *business plan* assessment criteria rather than sales training criteria employed after the fact. Moreover, text book analysis (Kotler et al 2002) more than adequately reviews the potential for multiple influences that may occur within a given DMU. Here the *business plan* is used to identify these influences and to accommodate the relevant details within any market audit.

5.5 Glossary of Terms & Chapter Summary

Listed below is an explanation of some of the terms used above if not already known to the readers. In addition to these terms the chapter is summarised by way of other marketing terms that may be used to assist with the audit.

Item	Marketing Term	Comments
1	Situation Review	A Marketing Audit is the initial phase of what is normally associated with a "Market overview" analysis. Additional research can be undertaken to strengthen the auditing process in terms of the items given below.
1.A	Market Mix	This term is used to define Products, Place, Price & Promotion.
1.B	Market Segmentation Studies	Market segmentation refers to the type of customers who may buy your product, the products they are looking for, the price they are prepared to pay as well as the environment that they exist within. It could be a singular segment in format where the product offering is suitable for your own geographic region. However, this could be split into further segments to include foreign customers from Spain for instance that may require different services to those from Germany. Marketing as a science provides substantial data in respect of market segmentation studies. Put simply it is a process that is employed that allows business owners to determine which segments offer the best opportunities for achieving or maximising company objectives.
1.C	Market Research	As the term suggests research can be a subject that attracts a high volume of data as well as the accumulation of substantial research time. Information can quite often be made available from national institutes, national industrial institutions as well as municipal offices. If further analysis is required it is suggested that contact be made with your local branch of the Chartered Institute of Marketing.
1.D	Gap Analysis	For those readers of an innovative disposition in an established business they should review what is available

		within the local environment. What is missing from this environment that could be provided as part of your product mix? This fact notwithstanding, the market audit will identify what gaps will exist within your marketing mix.
1.E	Product Life Cycles	Mostly associated with technology and commodity based products Product Life Cycles or PLCs impact upon all businesses. It is suggested that the PLCs associated with external factors as discussed further within the SWOT analysis chapter impact more upon product/service supplier analysis criteria. Here we can introduce another heading by way of technology lifecycles or TLCs.
1.F	Diffusion of Innovation	A complex statement for what is a relatively straight forward planning requirement. It considers the need to ensure that business staff, as well as customers fully understand the value of any innovation undertaken. This is particularly important for those companies where the majority of sales are attributable to buyers (market intermediaries) rather than users.
1.G	Business Development Matrix	It is normal for marketers within a business environment to provide a matrix (descriptive representation) that simplifies the investment process for the business. Such a matrix (adapted from Ansoff) is provided below. 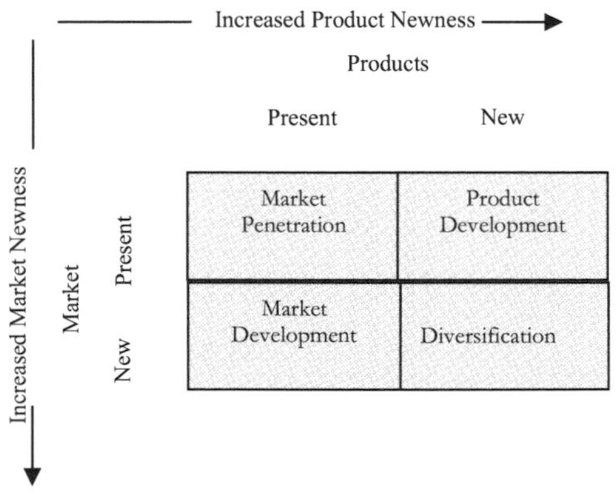 Increased Product Newness → Products Present / New Increased Market Newness ↓ Market: Present / New Market Penetration / Product Development Market Development / Diversification The main attraction of such a matrix is that it allows a business to assess how the factors given within the matrix interact with each other with the intent of directing

		investment in existing products and/or new products for this example.
1.H	Forecasting	Always a subject matter that attracts arguments as to the accuracy of any statistical data used, forecasting suffice it to say is not the easiest of subjects to discuss. Data may be available from an external source but may not include any specific data that may apply to the matrix provided in sub chapter 1G – if used. It is at this stage of the planning process that statistical data may be required from a greater geographical area undertaken by singular statistical authority.
1.I	Patterns of Behaviour	This statement considers the behavioural aspects of the target market selected. Here we have one of the stages of the planning process where one may need to revert back to the aspirations for the business in terms of product and/or service. Suffice it to say that the subject can attract substantial definition depending on the behavioural patterns associated with a given target market. There is the potential for this sub chapter to impact on the following sub chapter 1.J.
1.J	Unique Market Demands	It is often said that knowledge gives rise to power. If we can use this knowledge to identify unique market demands we will be able to maximise the opportunities for the business. Such a scenario must always be borne in mind. Remember, you can create something unique for your environment without relying upon the environment to create something for you. A cider manufacturing concern local to the author with an in-house pig-farm comes to mind for instance. Here the objective is to differentiate your business in a way that it makes potential visitors want to seek out your cider farm (in an area where there are many) rather than others.
1.K	Packaging	In what way should you package the products or services you have to offer? A question made much simpler if you are offering a boxed commodity such as a steam iron for example.

		Here there is a need to consider the packaging of your products that makes your product more attractive for your customers in a value creative manner. For instance lawn mowers now have large grass collection receptacles fitted to them. PVA paint revolutionised that particular industry. Furthermore, these same industries continually strive to expand the product mix in a way that reduces the effort associated with mowing a lawn or painting the lounge area. Your advertising material should always endeavour to state the value of each individual item of the package offered.

Table 5.5

The use of any information derived by the information gained from these tables will allow us to identify a market overview in six more concise statements, in particular the issues that will affect corporate objectives analysis and reasoning.

- Market Structure – who is that buys your product here in the UK? What export market exists?
- Market Trends – what sort of products are these potential customers seeking?
- Key Market Segments – how can we group together the needs of these potential customers?
- Segment Overlap – do some of these segments have similar needs?
- Gap Analysis – what additional products or services need to be provided that are not already available? Do you need to retain any specific information to boost long term strategic needs?

It will be noted however that these bullet points serve only to provide an analysis in relation to the factors that impact directly upon our marketing mix (Table 7B – Chapter 7.0). Without further marketing analysis the sales department will encounter barriers to orders that will directly affect turnover and the corresponding cash flow.

There is a need to incorporate within any survey data that will allow further analysis of certain specific factors that will have a direct or indirect influence on the purchase of the product or service and its consequent impact upon corporate objectives and strategy review.
Here the factors referred to are defined as PEST (Political, Economic, Social and Technological) factors. They form part of the information summary depicted as Tables 5.6 & 5.7. The manner in which they impact upon the "external audit" in relation to "the market" and the "competitor" base as an individual entity must always be observed.

For example high technology based consumer products may be priced such that they have limited markets owing to a low wage based political environment. This could suggest offshore manufacturing bases located within the appropriate country. Similarly, cognisance would need to be made in respect of alternative language usage in the form of operating the product and its corresponding manuals.

It is the influence imposed upon the "internal audit" assessment" caused by the external audit that will determine the overall impact of any ***business plan***. Where the data obtained is devoid of data sought there exists the opportunity for companies to create their own records.

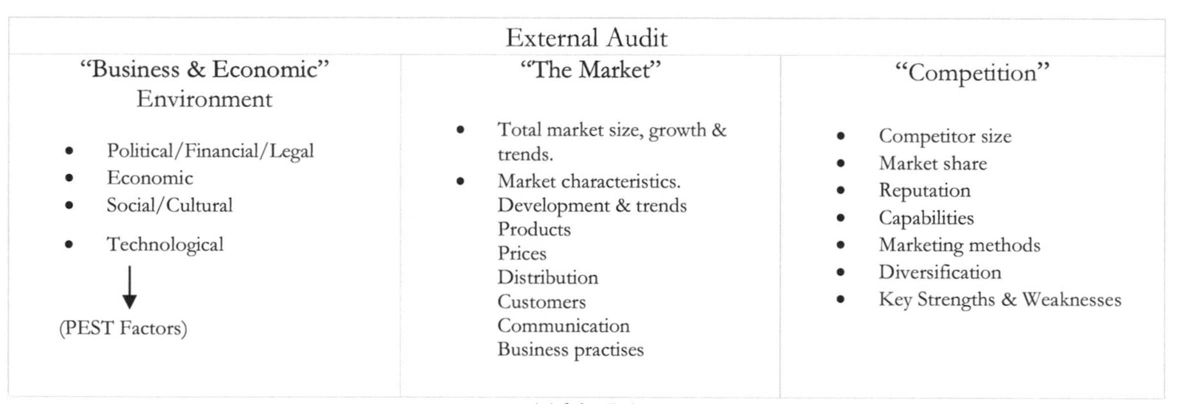

External Audit		
"Business & Economic" Environment	"The Market"	"Competition"
• Political/Financial/Legal • Economic • Social/Cultural • Technological ↓ (PEST Factors)	• Total market size, growth & trends. • Market characteristics. Development & trends Products Prices Distribution Customers Communication Business practises	• Competitor size • Market share • Reputation • Capabilities • Marketing methods • Diversification • Key Strengths & Weaknesses

Table 5.6

Internal Audit	
Marketing Operational Variables	Marketing Information
• Sales (total by geographical location, by guest type and their preferences) • Market share • Profit margins/costs • Marketing practise	• Market mix variables:- Product/service solutions Price Promotion

Table 5.7

Clearly, our marketing audit has many variables for consideration. Yes there is a need to maximise the amount of data we would wish to seek. However, any questionnaires used to derive this same data should it is suggested consider boundary definition criteria that may be associated with any given question. The reason for this will become more evident during the dissemination of the market information and its impact upon the remaining chapters.

For instance why accommodate all of the market needs into a target product or service. To this end, there will always be an option for multiple product variables (with optional extra cost) as well as the formation of a product family with anticipated Mark 1 or Mark 2 versions for later use.

5.6 Chapter Summary

The data derived within this chapter while important to the *business plan* will most probably be too great in volume to attach to the main part of the *business plan* submittal. Should this be so it is suggested that it forms the basis of an appendix used to support the *business plan.*
If wishing to include audit data it would be prudent to include only the information that will affect the strategies used to achieve the corporate objectives employed as discussed within the previous chapter.

Typical Example

Most companies may question the existence of contemporary market information within their organisation. For this example Eugene has given his approval to undertake a market audit. Manfred explained that IVS operated using statistics derived by way of SIC (Standard Industry Classification) and that the Department of Trade & Industry (DTI) should be contacted to gain the information she had requested.

Following a protracted period of correspondence exchange with the DTI and their own industry association Hayley had substantial data from which she could begin her audit. The data was most helpful by way of market users and associated volumes very much in the same way as the information used to describe our "where are we now scenario".
Indeed, she observed other potential target markets such as public buildings and the paper and pulp industries sectors. However she also noticed that the usage values given were based upon what the users bought. It was possible that these same values may not reflect what they specifically wanted and the impact this would have on the marketing mix and market value.

Manfred suggested that the possible reason for the latter scenario was because IVS had evolved into a "sales" based organisation and that their competitors operated on the same basis. He further stated that his work was now concentrated on meeting targets imposed by the sales channel.

He further admitted that their distributors now determined customer usage rates and the mix of product solution for several of their target markets. The audit undertaken by Hayley must consider operational needs using the existing sales channel and product mix but also take into consideration the long term strategic objectives they had discussed.

Hayley began the process of how her marketing plan could match the corporate objectives with what users bought compared with their needs arising out of the environment they operated within. To simplify this task she undertook an ontological assessment of their market environment as discussed further in chapter 9.1 hereof. Her observations are used as the basis of the first sheet associated with this part of the *business plan* as given in Fig 17. These are the "sales channels" that operate across the business macro environment; sometimes misused by way of referral to as "routes to market". Here it could be argued that "routes" reflect the macro environment arising out of the sales actions that must be undertaken for a "sales" based organisation to undertake trade. Whereas a "sales channel" is the appropriate conduit put in force to maximise the efficiency of a given sales process arising out of a balanced market solution in respect of the marketing audit.

The reason for this type of presentation at the outset of this chapter presentation is of a high a degree of importance. In the first instance it can be seen that certain customers may not in fact be the product users. To this end the market intermediaries purchase the product as part of their overall market offering. Similarly, architects are seen to influence the buying process both of whom are not the end users. Put simply we must understand the influence our sales channels will have when setting our audit questionnaire document.

Clearly certain questions will have more relevance to certain individuals compared to others. A demand for a lowest priced solution could be over ruled by a need to retrain many.
Conversely certain responses will have more relevance by way of their impact upon our importance in determining strengths, weaknesses, opportunities and threats. Such a scenario will impact upon the audit undertaken for each market segment hence its use as part of the *business plan* presentation.

Although a sheet should be provided for each target market this example considers only three markets for simplification purposes.

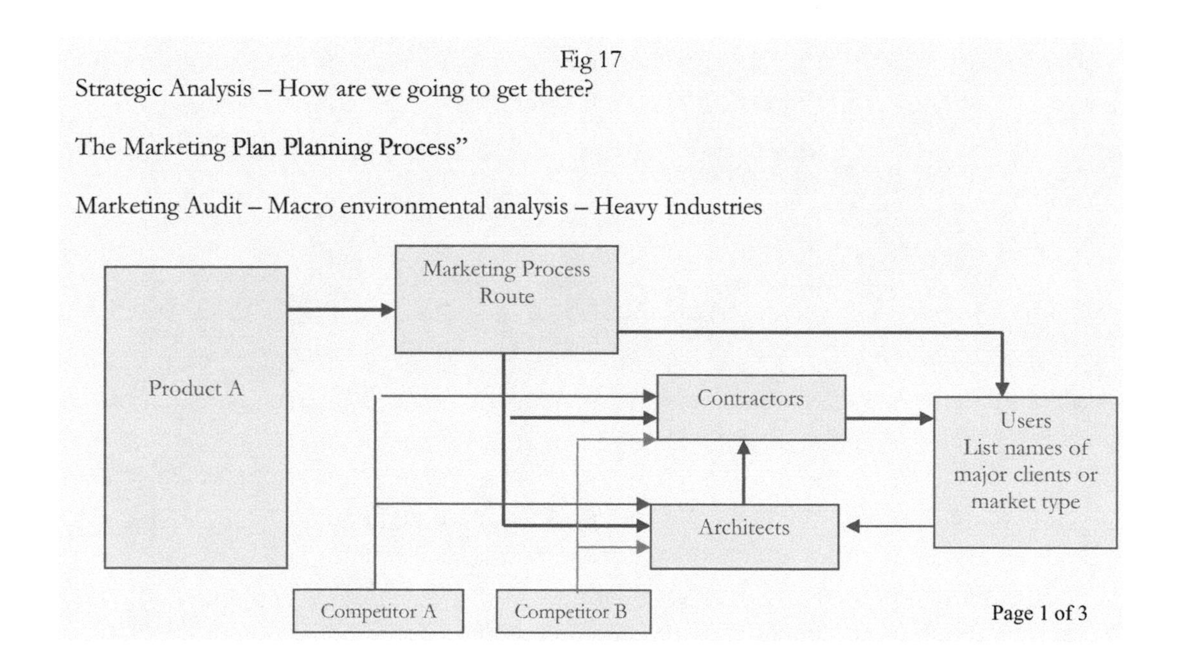

Fig 17
Strategic Analysis – How are we going to get there?

The Marketing Plan Planning Process"

Marketing Audit – Macro environmental analysis – Heavy Industries

From this sheet we can ascertain that our competitors do not endeavour to seek and supply accordingly the needs of the end users. Note also that if using an animated slide presentation to show this activity it is suggested that the competitor data be added in increments.

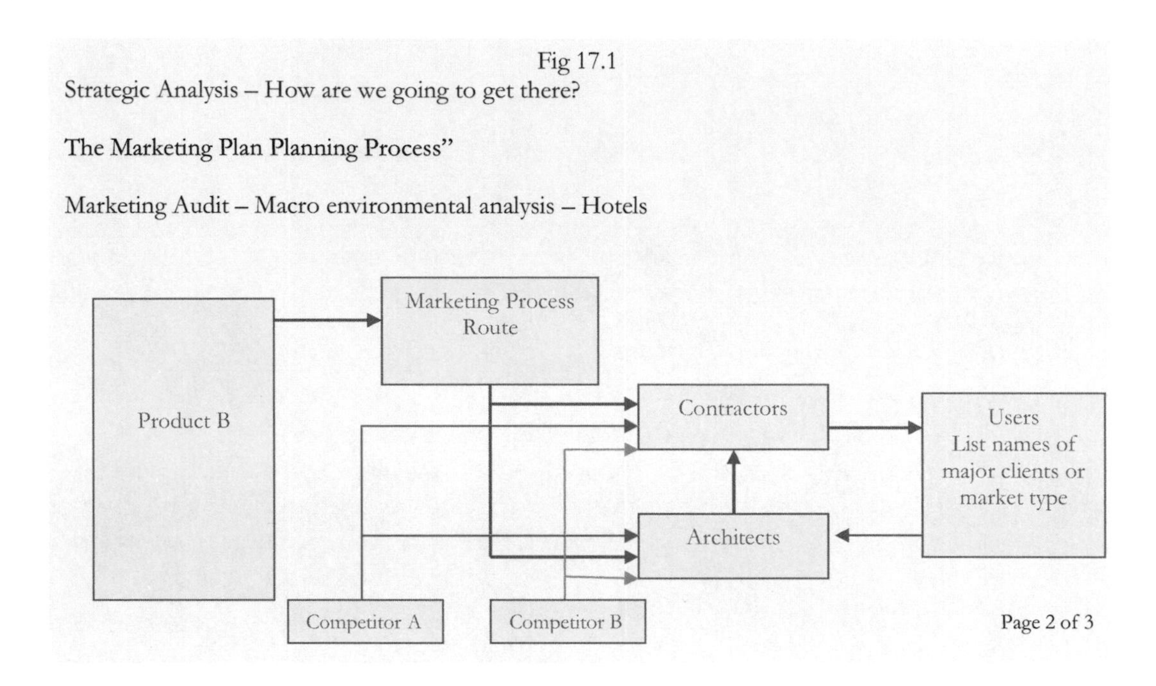

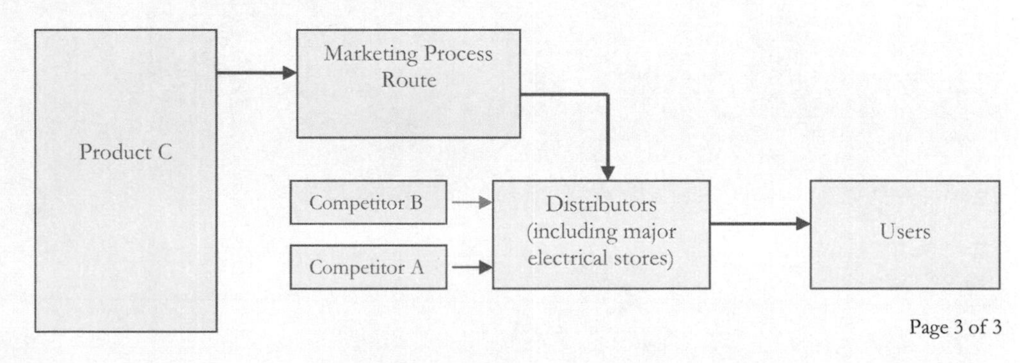

The major observation during this process is that any questionnaire associated with a market survey would require different areas of focus. To this end, there is a need to assess the process needs for the market intermediaries as well as any benefits expected or otherwise for the end users. Indeed, there would be questions for each micro environment that would not apply to both segments. Furthermore, the end users associated with the domestic market irrespective of product type would need some form of national assessment; a potential major task to say the least.

The metrics employed for the latter and the maxims (principles) associated with them has the potential to attract much criticism depending upon the research principles employed. An approximate cost for such a survey by a research company was suggested at £15,000.00. This gave rise to the use of the auditing principles put forward within this chapter together additional techniques given in table 5.8 below.

Market intermediaries:	To use the auditing principles given within this chapter to assess in what way will their use will improve IVS business activity? To use the auditing principles given within this chapter to assess what is it that influences non purchasing intermediaries (architects)? Survey sample: All major intermediaries within all target markets.
End users excluding the domestic market	To use the same techniques employed for the "market intermediaries" given above. Survey sample: all major users within all target markets.
Domestic Market	To use the same techniques employed for the "market intermediaries" given above. To undertake a sample survey within a selected region within Somerset. To prepare a formal enquiry for a formal market research following the completion of the audit suggested above.

Table 5.8

For the purpose of simplification the data provided hereafter is based upon a completed market questionnaire and data tabulation for the selected target markets as indicated. The questionnaire used combines all of that proposed within this chapter as an integral document but is summarised as part of the market audit Table 5.9 given below.

Product A "The market audit"– Market Intermediaries

Focus	Results
Customer needs and buying factors	• 90% were seeking a simplified tender and documentation process. • 95% mandated constant reports on the delivery process. • 85% would wish to implement virtual contract management techniques.

	• 100% wanted a more flexible commercial offering. • 100% wanted fit for purpose plus optional tenders. • 100% wanted a more defined scope of supply. • 90% wanted discounts for stage payments.
Products and customers	• 60% suggested a greater mix of heating & ventilation power output for the Food & Beverage industry. • 70% suggested a greater mix of heating & ventilation power output for the Steel industry. • 90% suggested a greater mix of heating & ventilation power output for the Public Buildings industry. • 70% were entirely happy with the electrical controls provided. • 30% would like a more simplified control for the Public Buildings industry. • 30% would like some form of wireless control for all industries. • 100% thought that all suppliers' products made too much noise.
Key products	1. 95% preferred to operate on a turn key basis. 2. 80% preferred to offer long term maintenance contracts. 3. 90% preferred a site survey prior to confirming any contract. 4. 90% were insistent upon a complex documentation package. Given the high percentages listed in relation to those surveyed no separate equivalent market potential is listed within this audit factor. For example item 1 may represent 97% of the market by the market value they represent.
Marketing priorities and critical success factors	1. 80% wanted 3 extra heating outputs for the Food & Beverage industry. 2. 90% wanted 2 extra ventilation outputs for all markets except Steel industry 3. 10% wanted 3 extra ventilation outputs for the Public Buildings industry 4. 60% would immediately implement wireless control. 5. 100% would immediately implement effective noise reduction schemes. This audit factor is listed as differentiating factors in addition to the factors already provided by the market intermediaries.
Market segmentation	The methodology employed to seek out potential markets suggested a

redefinition of the market sectors provided within the *where are we now?* subject previously alluded to.

Segmentation Techniques

Separate pie charts that further defined existing data were produced. To this end, heavy industries now comprised the steel sector, mining and the petroleum industries. A total value was given for each industry and where the potential business existed as given hereunder. Note, the values provided reflect total industry spend. IVS by experience are aware of the approximate percentage of this spend that equates to heating & ventilation.

Steel sector: £20M UK / £400M Overseas (break down into region)
Mining sector £15M UK / £800M Overseas (break down into region)
Petroleum sector £100M UK / £5 billion Overseas (break down into region)

Light industries and the domestic markets were similarly defined although not stated for simplification purposes.

Matching Criteria

Here the audit focused upon the generic market needs perceived through experience and market feedback as well as bespoke needs derived on the same basis.

A separate table was produced that defined matching criteria as well as any indifference for each segment.

The heavy industrial segment is used as an example.

Activity	Steel	Mining	Petroleum
Electrical Specifications	Common	Common	Common plus hazardous needs
Mechanical Specifications	Common	Common	Common plus hazardous needs
Documentation	Common	Common	Common

<table>
<tr><td></td><td><table><tr><td>Turnkey Based Project Activity</td><td>Common</td><td>Common</td><td>Common</td></tr></table></td></tr>
<tr><td>Company priorities</td><td>• 100% state existing manufacturing facilities considered superior to competitors across all segments.
• 100% state existing product mix is considered superior to competitors for all segments except domestic (0%)
• 100% would cooperate in an existing process review Vs existing product mix for Heavy & Light industries.
• 100% believe the advent of global warming will increase the need for more ventilation across all segments.
• 100% believe the advent of global warming will increase the need for innovation within the industry across all segments.</td></tr>
<tr><td>Market sizing and shares</td><td>Heavy & Light Industries
All market intermediaries state that UK growth over the next 5 years is static but expect significant growth (20% p.a.) for the overseas markets during the same period.
Domestic Industry
All market intermediaries anticipate high growth over the next 5 years.</td></tr>
<tr><td>Life cycle and competitive position</td><td>Heavy & Light Industries
All market intermediaries believe existing technologies employed are suitable for any specific retrofitting arising out of any new product development needs for all suppliers.
Domestic Industry
All market intermediaries believe that product life cycles will continue to be of a short duration owing to globalisation and technological enhancement.</td></tr>
<tr><td>Competitors</td><td>The data obtained was produced from a combination of market feedback data together with that secured from the web sites of Competitors A&B.

Heavy & Light Industries
• All market intermediaries believe that the existing IVS market growth (potential) will be affected by the growth in Competitor A&B product/service mix. Competitor price is the dominant factor for Products B & C.</td></tr>
</table>

	• No one has 3 extra heating outputs, increase in UK market size is £1M/year • Only Competitor A has 2 extra ventilation outputs, UK market size is £2M/year. • No one has 3 extra ventilation outputs, increase in UK market size is £1.5M/year • Competitor B undertakes long term maintenance contracts but only for their products. • Competitors A&B facilities are located within 50mile radius from 90% of all market intermediary offices. • Competitors A&B have neither the facilities nor manpower to accommodate the overseas potential available. *Domestic Industries* • Both Competitor A&B source standard products that are manufactured in China with logistics that allow delivery direct to the market intermediary despatch depots. • All market intermediaries state that they operate within a technological environment that is highly turbulent. Innovation can sometimes create a scenario where demand outstrips supply. Generic • Current supply meets market demands for all market segments ergo no new market entrants are anticipated arising out of current market factors. • Indirect factors such as governmental edicts/initiatives associated with global warming can give rise to new technologies and new market entrants.
Market environment	• The market accepts that IVS are a leading brand for both the heavy and light industries and that it is this image that allows sales to be achieved within the domestic sectors. • All market intermediaries would prefer that IVS retain an image as heating & ventilation specialists rather than operating as separate

<table>
<tr><td></td><td>entities.
• End users (Heavy & Light Industrial) would prefer major project activity by major contractors but mandate preferred product supplier status.
• The domestic intermediaries do not care about brand image unless it is mandated by the end users.
• All market intermediaries suggest an update of IVS corporate vision to maintain its brand image (based on competitor growth).
• All market intermediaries believe global warming will impact upon company health and safety regulation change.</td></tr>
<tr><td></td><td>THE REMAINING FOUR AUDITING FACTORS FORM PART OF THE BUSINESS PLAN PRESENTATION</td></tr>
<tr><td>Market summary</td><td>Limited data is provided for simplification purposes. Readers may wish to expand upon that given to meet and end to their aims.

Strategic Analysis – Where do we want to be?
The Marketing Plan Planning Process"
"Market Segments"
Page 1 of 7

Strategic Analysis – Where do we want to be?
The Marketing Plan Planning Process"
"Market Segments – Selection Criteria"
• Sector market value & growth.
• Balanced & matched market needs in respect of IVS product mix.
• Limited market expansion in terms of mechanical enhancement.
• IVS ability to meet technological enhancement requirements.
Page 2 of 7</td></tr>
</table>

Strategic Analysis – Where do we want to be?

The Marketing Plan Planning Process"

"Market Segments – Heavy Industry"

Market Segment	Value (P.A) UK	5 yr P.A. Growth	Value (P.A) Overseas	5 yr P.A. Growth
Steel Sector				
Food & Beverage				
Paper & Pulp				
Petrochemical				
Oil & Gas				

Insert appropriate values

Page 3 of 7

Strategic Analysis – Where do we want to be?

The Marketing Plan Planning Process"

"Market Segments – Available Market - UK"

Market Segment	IVS	Competitor A	Competitor B
Steel Sector			
Food & Beverage			
Paper & Pulp			
Petrochemical			
Oil & Gas			

Insert appropriate values

Page 4 of 7

The following presentation should be undertaken for each product in relation to individual segment sector needs.

Strategic Analysis – Where do we want to be?

The Marketing Plan Planning Process"

"Market Segments – Product Compatibility - UK"

Industry Needs	IVS Product A Features	Competitor A	Competitor B
Feature A			
Feature B			
Feature C			
Feature D			
Feature E			
Feature F			
Feature G			
Feature H			

State yes or no when referring to the competitors or provide actual feature rating or output. State where IVS must undertake NPD.

Page 5 of 7

A summary sheet can then be undertaken that defines how the product meets the overall needs of the market segment. Clearly, this exercise should be done for each product type in relation to a specific market segment.

Strategic Analysis – Where do we want to be?

The Marketing Plan Planning Process"

"Market Segments – Product Compatibility - UK"

Product A Features	Steel	Food & beverage	Paper & Pulp	Petrochemical
Feature A				
Feature B				
Feature C				
Feature D				
Feature E				
Feature F				
Feature G				
Feature H				

Combine the previous sheet for all market sectors and undertake the same exercise for all product types.

Page 6 of 7

Concurrently there is a need to analyse the data for the domestic market secured using the principles outlined in sub chapters 5.1A given within this chapter. The main criterion observed were:-

- Products could not be compared on a like for like basis with competing suppliers.
- Loyalty status was high amongst high earners.
- High earners would purchase ventilation systems if low end solutions were to be made available.
- Flexibility for portable heating and ventilation products is a dominant factor.
- Easy of installation is a dominant factor for fixed installations.
- Simple controls are a dominant factor for all types of installations.

From the sheets put forward we should be able to identify the gaps in product terms in relation to market needs that the audit has revealed.

Strategic Analysis – Where do we want to be?

The Marketing Plan Planning Process"

"Market Segments – Market Positioning"

- Retain market leadership for the industrial sectors by enhancing the IVS product mix to limit market gaps.
- To balance IVS sales processes in relation to market intermediary and end user process enhancements.
- To maintain the domestic market brand image but to maximise sales through innovation.
- To use this same innovation to offset the impact of global warming

Page 7 of 7

Strategic Analysis – Where do we want to be?

The Marketing Plan Planning Process"

"Market Segments – Market Priorities"

- To assess off shore manufacturing for commodity based components for the domestic market segment product mix.
- Assess innovation in terms of ROCE and resource needs relative to market positioning needs.
- To analyse the ROCE and resource needs associated with product maintenance market potential.
- To maintain the domestic market brand image but to maximise sales through innovation.

Page 1 of 1

Strategic Analysis – Where do we want to be?

The Marketing Plan Planning Process"

"Critical Success Factors"

- IVS must introduce a proactive market sensing system.
- IVS must re evaluate its human and manufacturing resource capabilities to meet UK expansion as well as potential overseas business.
- IVS must evaluate the level of any investment needed to meet customer priorities.
- IVS must reassess its management structure in terms of resource and/or any additional training needs.

Page 1 of 1

Strategic Analysis – Where do we want to be?

The Marketing Plan Planning Process"

"Marketing Objectives"

"There are two separate marketing objectives for the company. One that is common for all industrial sectors and a separate objective that will be employed for the Domestic industry"

Page 1 of 3

The following two sheets are used to describe how the marketing objectives as part of the overall *business plan* will allow the corporate strategies given within the previous chapter to be implemented.

Strategic Analysis – Where do we want to be?

The Marketing Plan Planning Process"

"Marketing Objectives Industrial Sectors"

- The external factors that compliment the market audit used to create this *business plan* indicate that IVS corporate strategy is fully achievable.

- The marketing plan forming the basis of this *business plan* will be used to identify new product development and any innovation associated with the variables not currently under consideration by IVS that are needed to achieve short and long term objective needs.

- Internal factors are similarly disposed provided that the variables and boundary definition criteria associated with corporate strategy policy and programmes form part of this *business plan* as given hereof.

Page 2 of 3

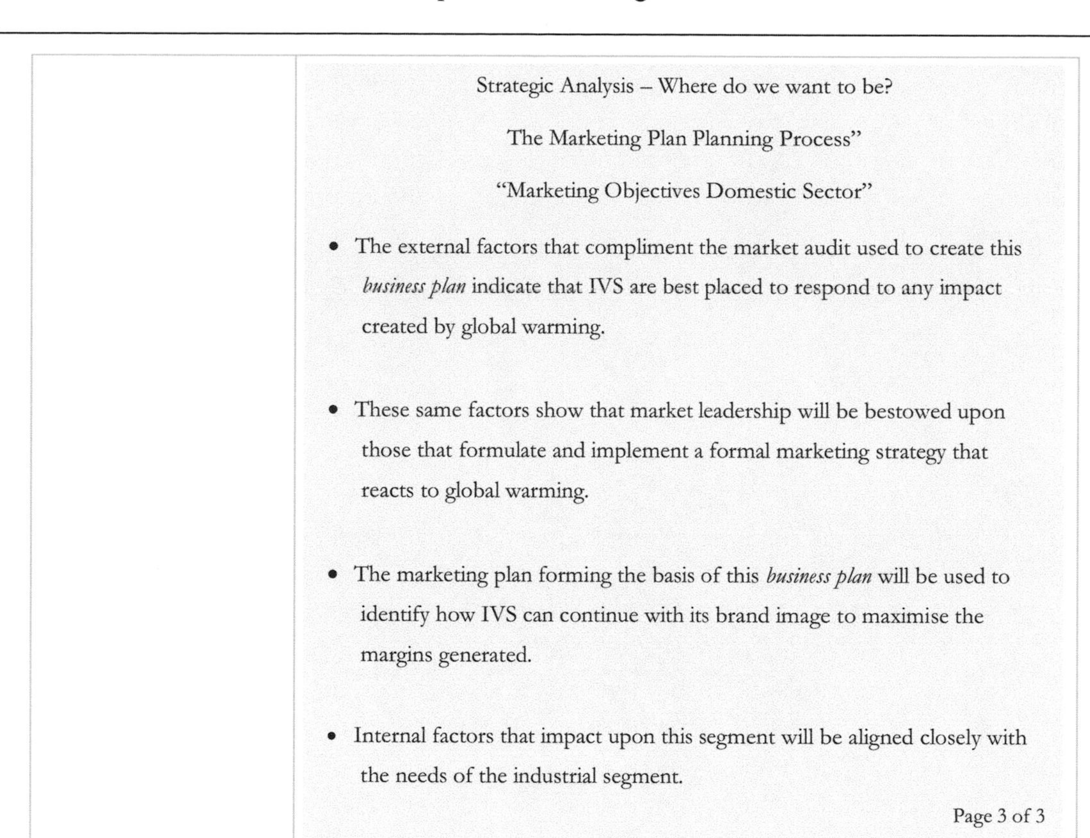

Strategic Analysis – Where do we want to be?

The Marketing Plan Planning Process"

"Marketing Objectives Domestic Sector"

- The external factors that compliment the market audit used to create this *business plan* indicate that IVS are best placed to respond to any impact created by global warming.

- These same factors show that market leadership will be bestowed upon those that formulate and implement a formal marketing strategy that reacts to global warming.

- The marketing plan forming the basis of this *business plan* will be used to identify how IVS can continue with its brand image to maximise the margins generated.

- Internal factors that impact upon this segment will be aligned closely with the needs of the industrial segment.

Page 3 of 3

Table 5.9

Both of the latter two sheets make reference to external and internal factors. These were discussed during the concluding part of this chapter as given above.

In most cases an audit as undertaken with the example provided will have answered most of the questions that will allow these factors to be identified in terms of how do they relate to IVS and the macro environment it operates within? Clearly, the iterative needs of the marketing plan must still be pursued by way of that suggested in Table 5.10.

External Factor	Internal Factor
Political & Legal environmental factors	How will any new product development and any innovation impact upon these factors? It follows that any regulation arising out of these factors must be assessed in terms of boundary definition criteria.

Marketing Operational Variables	Marketing Plan Function
Sales, Market share, profit margins, marketing practise?	The external audit undertaken defines what operational variables is achievable using existing or an expansion of the existing resource and network. What is not known are how any opportunities can be maximised and what risks apply and what market mix variables should be applied?

Table 5.10

These two concluding tables provide us with a final sheet for presentation purposes as given in Fig 18.

Fig 18

Strategic Analysis – How are we going to get there?

The Marketing Plan Planning Process"

"Marketing Objectives Synopsis"

"To undertake a SWOT analysis for IVS and to formulate and implement a corporate strategy that maximises opportunities and minimises risks"

Page 1 of 1

There will be those of you wondering what about the market communications audit? Yes this audit impacts upon our marketing and business plan strategy formulation but an over view of this same audit will reveal that it impacts upon people, their influence and any messages we may wish to convey. As a consequence, the audit data obtained will be kept in abeyance until we begin to formulate our strategies – chapter 7.0.

Using a hypothetical situation we can suggest that Hayley who is not an engineer has had difficulty with some of her colleagues who have used technological reasons as a barrier to audit data preparation and information dissemination. However, the information provided within the supporting criteria given below can be used to offset any such barriers. It will be noted that the comments provided will impact very much on the next two chapters.

Supporting business plan research criteria

The need for a SWOT analysis and its usage is well documented. In terms of its possible relevancy associated with this book is the impact that technology has in terms of "opportunities" and "threats" and their impact upon a market segment as suggested by McDonald, M (2003, P.578). A market segment can then be broken down into "customer orientated" and internal SWOT analyses suggests Piercy, N.F (2003, P.543).
Here he considers that "technical expertise" (possibly considered a strength) in the larger companies could imply bureaucracy which, itself is a weakness. He also states that experience could be considered strength but that this could imply "no innovation" techniques.

Many markets would appear to be dominated by technical specifications that are generic to the industry. These same specifications are used to identify the minimum needs of a product to undertake a specific technical function. The validity of these specifications is also controlled by external organisations such as the IEC and ENA approvals panel. They are prepared by client engineers and are then assessed by engineering personnel within supplier organisations before any formal development of a product or quotation is undertaken.
This may in some way account for some of the failures of technology that have occurred where manufacturers have perceived product differences but where customers could not as observed by Cahill, D.J & Warshawsky, R.M (1993, P.19). They provide us with an analogy where the speed of a product operation was increased by nano seconds but what benefit did it create?

Technological marketing techniques would ensure that these same specifications would be given additional exposure across the segment as a whole. They could then be assessed as part of the SWOT analysis process in terms of the ability to create a focus of competitiveness. Any technical advances that may be required to remain competitive will require marketing as a subject to transform the technical advantage into competitive advantage as stated by Dhanani, S et al (1997 P.160).
This would suggest that any barriers stated exist only because of a lack of market analysis in relation to totally integrated market needs (Piercy (2003, P.238)). This further supports but also expands upon a need to consider internal and external influences (Meldrum (1994)).

The need to assess market segments in relation to micro and macro environmental focus will serve to respond to the needs of suppliers, products and markets (Meldrum (1995, P.48)), he states;
"are closer to the heart of what marketing is about"

Where he suggests the use of "Internal Issues" and "External Issues" for consideration the external issues put forward do not consider the impact upon the client/environment. Nor is a direct relationship with this impact and the impact that the environment imposes upon the microenvironment as suggested by and its impact upon questionnaire content.

It is their impact upon the methodology employed to focus on the questionnaire to be issued that is seen to be of importance as shown in Tables 5.11 & 5.12.

Author/s	Macro environmental impact upon the microenvironment
Brooksbank (1999, P.78)	Suggests that a company should employ a marketing orientation that concentrates on satisfying its customers using an ever-evolving process. It should not concentrate on producing a technically orientated product at the lowest cost.
Goldsmith (1999, P.179)	Considers the advent of services marketing where "the procedures by which buyers acquire and use the product" as having an impact upon marketing mix strategies.
Simkin and Dibb (1998, P.409)	Analyse the importance of government regulations and sales as "market attractiveness" criteria
Zineldin (2000, P.9).	States that companies who do not adapt to changing technological conditions will face "painful competition"

Table 5.11

Author/s	Technological impact upon Internal Issues
Coates and Robinson (1995, P.14)	Expand upon Wing and Montaguti where they show that: "new product ideas are most likely to be more successful when R&D and marketing/distribution/sales or customers are involved"
Crick and Jones (1999, P.162)	Consider the match between product and its technology match with the market. They go on to argue for an R&D and Marketing interface within the marketing process in order to create this match.
Maile and Bialik (1998, P.53)	Expand upon Meldrum theories with their model for new product selection that not only aligns with customer needs but one that is profitable.
Meldrum (1995, P.49)	Suggests that the technology associated with a product and the product it produces must be undertaken in a way that aligns with the needs of the market it is targeting
Nijssen and Lieshout (1995, P.28).	Expand upon Coates and Robinson using awareness techniques of potentially available models for use with new product development.
Wang and Montaguti (2002, P.82).	The successful co-operation between R&D and marketing is indicative of the more successful firms. They show a definite link

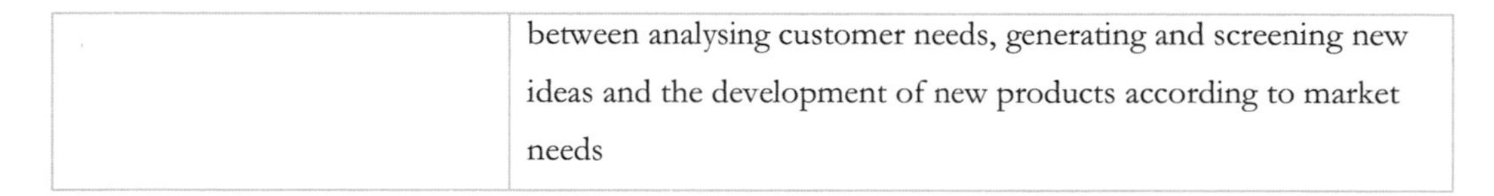

	between analysing customer needs, generating and screening new ideas and the development of new products according to market needs

Table 5.12

SWOT Analysis

Very much a business as well as a marketing term SWOT is an acronym for Strengths, Weaknesses, Opportunities and Threats. This part of the *business plan* is where we undertake to establish a realistic and objective appraisal of internal company strengths and weaknesses in relation to any potential external opportunities and threats in other words a SWOT analysis.
The latter is dependent upon the output of the marketing audit. That is it draws from the data established within the previous chapter. In business terms there is a need for any business to deploy its resources and that which it values in a manner that will achieve maximum return for its investment.

There are several ways that a SWOT analysis can be undertaken. In the first instance it is important to note that such an analysis should be undertaken for each target market segment. One such method is to employ what are known as "Motherhood Statements" (McDonald 2003) at the outset. A typical example is provided as given below but as can be seen, what could be listed as strength could give rise to a weakness. For instance a restaurant may specialise in gourmet food. While this could be seen a strength by the business owners it may reflect a weakness within the target market selection criteria. For instance the ratio of standard meals needs to gourmet meals needs for that area may be 5:1. Put simply, the gourmet food establishment targets only 20% of the market.

"We are an old established business

Strengths	Weaknesses
Stable	Inflexible
Trustworthy	Old-fashioned
Experienced	No innovation

'We are a large Business Group'

Strengths	Weaknesses
Comprehensive product range	Bureaucratic
Expertise	Offhand with customers
High status/stability reassures customer	No continuity of personal contact

It is important to note that the SWOT analysis will eventually form part of the strategic intent of the business (next chapter) and as such any motherhood statements considered by business owners or managers must have a real message associated with them. For instance we are an old established firm could be understood as we have always been in the business but so what?

Following this initial analysis and understanding there is a need to expand upon the same in the context of the external opportunities and threats that exist within the market segments. Here one of our main aims is to alert product designers and/or innovators to the necessity to create a balance between market wants and needs and how the information gleaned can be used to outperform competitive activity.

This particular research can be expressed in the form of a statement of the business and economic environment as well of that of the business under consideration as shown in Table 6.1.

External (opportunities & threats)	
Economic	Factors that impact upon the business are inflation, energy prices, and market unpredictability in terms of reduction in economic conditions as well as improvements in economic conditions. For instance, increases in energy prices may give rise to a review of different technologies associated with the distribution of energy as well as the manner in which it is employed. However, for every opportunity listed it is important to identify what threats could impact upon the same opportunity.
Political/Financial/Legal	Factors that impact upon the business are human rights needs, taxation, regulatory constraints such as advertising, product quality and trade practises. Here a threat could result in an opportunity. For instance how do health and safety impact upon a company's insurance premiums? Ergo, can your company produce a product that meets H&S requirements to an extent whereby it will reduce target market premiums?
Social/Cultural (Demographic)	The factors that impact upon the business are education, immigration, and religion and population distribution as well as family wealth (dual incomes) and an ageing population. The latter certainly has an impact upon the geographical region within which the author lives. To this end, there are numerous electrically operated vehicles to be seen day in day out.
Technological	The factors that impact upon the business are market needs versus business wants and the foregoing factors that impact upon technological application. Here the criterion is to limit any technological turbulence in respect of the latter factors. For instance Nuclear energy is known to be a low cost method of producing electricity but what its impact upon the

	environment and the political influences associated with it.
Internal (strengths & weaknesses)	
Potential sales	As determined by the marketing audit. Do you want to be a market leader or be considered as one of the others? Remember that most businesses have an eclectic mix of services they are able to offer and one need not be a leader in all of the potential services. Determine which factors (critical) impact most upon the business.
Potential market share	As determined by the marketing audit. You may already be a leader of services within a given area/s but may wish to expand but how? Similarly, do you want to be a leader in the additional services you want to offer? Here there is a need to bear in mind the skill levels that may be required to provide any additional or an expansion of existing services.
Potential profit margin	As determined by the marketing audit. Bear in mind the number of services you wish to offer and the cost of those services as they may impact upon the profit margin.
Marketing activity	As determined by the marketing strategy. Additional services may increase the cost of your marketing budget. They may also require additional management skills as well as time allocation.
Marketing mix variables	Market research, Product/Service development & range, quality, pricing package, sales aids, advertising, promotion and public relations. Each will create strength for the business but will most certainly give rise to some form of weakness. R&D is a prime example owing to its nominal cost and ROCE implications.

Table 6.1

It is to be noted that the variables stated may be expanded by those considered important by readers but not acknowledged hereof. This fact notwithstanding business owners will need to consider these same variables in a strategic format. That is, how do they impact upon the business and business management? Furthermore how should we understand and use this knowledge in a manner that will maximise our opportunities? Here a further model (that which is apparent) is provided for guidance as given in Fig 19.

	Internal	External
Good Points	Strengths What are we good at relative to competitors?	Opportunities What changes are creating new options for us?
Danger Points	Weaknesses What are we bad at relative to competitors?	Threats What emerging dangers must we avoid or counter?

Fig. 19 Hooley Et Al SWOT Analysis

As stated previously it is the external issues that are the sources of all the company's opportunities and threats which make an understanding of this model somewhat important. The benefits associated with its usage can be gleaned in several ways. It could be used as an *aide memoir* for our non-commercially based product design team and its eventual impact upon sales force selling techniques. The model presented as Fig 20 has more strategic implications. It is to be noted that this same model has additional usage in the context of marketing plan implementation as given in chapter 9. Furthermore, if we examine more closer the differences between these two models (Fig 19 & Fig 20) there exists the potential to determine resource skills gap between strategic needs and current assets. Thus any training needs can be identified and assessed if a specific competence does not exist currently.

	Opportunities	Threats
Strengths	Exploit existing strengths in areas of opportunity	Use existing strengths to counter threats
Weaknesses	Build new strengths first to take advantage of opportunities	Build new strengths to counter threats

Fig. 20 Hooley Et Al SWOT strategic implications

Clearly, some form of balance must be observed depending upon the type of industry one operates within. To this end, "Use existing strengths to counter threats" may only apply if the industry in question is maximising the technology stocks available at any given time for a given market. Where the market exists through expanding existing stocks (micro processors for example) resource application may be best suited to "exploiting existing strengths in areas of opportunity".

So how can we summarise these analyses to the marketing audit in order to create objectives (and strategies) based on that given within the marketing audit?
Firstly it is important to note that a marketing or *business plan* is not an essential requirement to undertake a SWOT analysis. It is a marketing function that can be undertaken in conjunction with a marketing audit separate to the other chapters discussed hereof. Put simply the analyses will provided a structured method of auditing any existing business activity.

This may suit many businesses as it is unlikely that a marketing plan is in force.
Irrespectively, there is a need to maximise the information associated with the external factors. Text book data is available that simplifies this process. For each market segment type it is suggested that additional research as given in Table 6.2 be undertaken.

Step	Comment
Step 1	List the principal opportunities (and where they exist) determined by the market. In particular refer to any statistical data provided by any national statistics authority as given within the marketing audit chapter.
Step 2	Allocate a code to each of these (e.g. A, B, C etc.). For instance, a hotel chain could list: Sporting groups (by type of sports activity), a mix of age groups, short stay holidays. Long stay holidays, business accommodation (by business profession e.g. doctors, teachers etc), recreational (hikers) and so forth.
Step 3	Allocate a number between 1 and 9 to each code. The number 1 means that in your view there is little chance of that group requiring your services within the planning timescale (say 3 years). A 9 would mean that there is a high probability of it occurring within the planning timescale. For instance, if you are a hotel located near a newly built highly publicised sports stadium but currently have facilities that cater only for say commercial travellers the marketing audit would suggest many more opportunities exist.
Step 4	Allocate a number between 1 and 9 to indicate the importance of the impact each of these opportunities would have on your business were it to occur. Check any statistical data that exists both locally and nationally.
Step 5	Now put each of your opportunities on the opportunities matrix (Fig 21 below).
Step 6	You will now have a number of points of intersection which should correspond to your

	coding system.
Step 7	All those in the top left box should be tackled in your marketing objectives and should appear in your SWOT analysis. All the others, while they should not be ignored, are obviously less urgent. The whole exercise should now be repeated for an equivalent "threats matrix.

Table 6.2 Adapted from McDonald (2003)

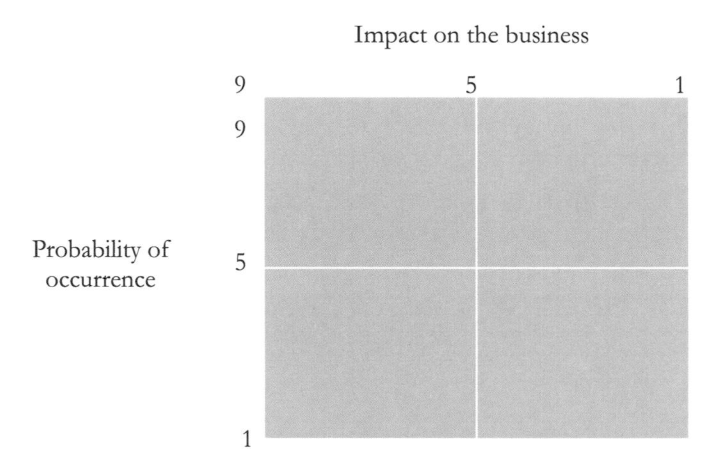

Fig 21 – Opportunities matrix (McDonald 2003)

6.1 Chapter Summary

A successful SWOT analysis if undertaken correctly should align with text book definition criteria namely:-

> "The organisation can begin to see where its strengths might be best deployed, both offensively and defensively, as well as where its weaknesses leave it vulnerable to market change or competitive action."
> Source: Davidson, H (1997) *Even More Offensive Marketing.*

The conclusion can be used to describe the outcome of a SWOT analysis irrespective of the methods employed to structure the *business plan.* It can apply to a "Sales" based organisation as well as a "Marketing" based company. Albeit it is considered unlikely that sales based organisations will have access to the tools needed to adapt the SWOT analysis to their *business plan.* Here it could be argued that we can again refer to a paradoxical link to company intent and its ability to deliver it. For instance, technological boffins may be employed to create a product or service offering but there may be difficulties in manufacturing (quality issues?) as well as an inability by non technological sales staff to effectively undertake the selling process.

Chapter example

As with the previous chapters there is a need to undertake a series of discussions to arrive at that which should form part of the *business plan.* The chronological process described refers to the need to assess at the outset the external opportunities and threats in the form of PEST factors (Political, Economic, Sociological, Technological). Although slightly expanded upon hereof PEST factors is an often used acronym by marketers.
The information used to complete this example (Table 6.3) is borne out of fact derived from the market audit and by information gleaned from the internet. Supporting text is provided at the end of this example.

External factor	Opportunity	Threats
Economic		
• Energy prices are extremely turbulent.	• Global warming would suggest less energy usage for heating purposes.	• Increase in ventilation product output requirements as required by the market would suggest the use of more electrical energy.
Political		
• Current political aims consider a global reduction in green house gases.	• Our existing product mix is rated the most energy efficient.	• An increase in product output will require a higher volume of refrigerant.
• New safety laws are being introduced to govern the disposal of refrigerant.	• The new refrigerant employed by IVS is reusable.	• Reclamation logistics are unknown.
Financial		
• The potential exists for government funding (local or otherwise) to convert old large ventilation systems that do not comply with new regulations.	• This type of business would support very much overhead recovery during periods of low industrial activity.	• Boundary definition criteria associated with such promotions could create complex contracts.
Legal		
• Insurance surveyors are considering revising their mandates for heating & ventilation minimum specifications.	• IVS are best positioned to influence the end users and its congruent impact upon market intermediaries and architects.	• Contract legal and documentation needs are an unknown factor.

Social/Cultural • Global warming would create more hotel/office buildings and domestic ventilation expenditure.	• IVS overall global warming corporate vision and its intended brand association will create more definitive long term business opportunities.	• We currently are unable to compete on price for those users other than brand loyal users. Furthermore, there exists a difference between low end domestic user needs and those of low end industrial users.
Technological • Existing technology usage world wide is based upon the provision of a solution to create a means to and end.	• IVS business plan would suggest the application of technology based upon segment needs and associated overlapping segment needs.	• IVS will need to invest in technology life cycle research rather than product life cycle evaluation.

Table 6.3

We next review our strengths and weaknesses in relation to the internal factors; Table 6.4 refers:

Internal Factor	Strength	Weakness
Potential sales • Industrial segments	• Originally close to end users thus creating more value and market leadership.	• Impetus has lapsed thus the increase in competitor share.
• Domestic segments	• Brand value sustains high end earner sales.	• Competitor image is improving.
Potential market share • Industrial segment	• IVS can retain and grow market leadership with NPD and/or the use of innovation.	• Existing management structure does not align with business growth.
• Domestic segment	• IVS already has a brand image but is adaptable to overseas venture participation to create lower priced products and to adapt new ratings.	• The costs and logistics associated with overseas manufacture have yet to be reviewed.
Potential profit margin • Industrial & Domestic segment	• IVS corporate objectives allow an increase in product mix arising out of market	• IVS management structure is not yet in place to support growth arising out of the

	needs and an increase in service needs arising out of the external factors.	audit.
Marketing activity • Industrial & Domestic segment	• Past experience shows that the marketing undertaken by Manfred & Eugene at the outset was proven to be successful.	• An expansion of IVS market mix may require formal marketing expertise which the company does not have.
Marketing mix variables a. market research	a. Audit suggests IVS is the only market orientated company across all industries.	a. Data is based on historical experience. IVS is now sales orientated.
b. product mix	b. IVS has the most usable product mix (Industrial)	b. Same comment does not apply to the domestic segment.
c. service mix	c. Technology is available to provide an excellent product.	c. IVS operate on a reactive basis only.
d. quality	d. IVS is shown to provide the best quality (Industrial and domestic high end earners)	d. Majority of domestic market will not pay for high quality.
e. reliability	e. IVS is shown to provide the most reliable solution (Industrial and domestic high end earners)	e. Majority of domestic market will not pay for high reliability.
f. presentations	f. Eugene & Manfred both visit customer premises.	f. Visits are undertaken on a reactive basis.
g. public relations	g. Manfred & Eugene often meet influential clients outside of business hours.	g. Feedback is known only to Eugene & Manfred.
h. after sales	h. After sales is shown to be a profitable concern.	h. All work is undertaken on a reactive basis.

Table 6.4

These two tables have the potential to be expanded upon depending upon the extent of the audit undertaken and the response criteria provided. Our use of that given and its congruent impact upon the marketing plan/ *business plan* as stated within this chapter is of paramount importance for setting our objectives and strategies.

These it must be said are focal to all our business aims. It will define objective and strategic criteria associated with the product mix we must choose, how we produce them and to whom we can maximise our potential. Let us also not forget what impact upon the market we can create with our competence/ talent.

One method of expanding upon that given is to analyse strengths, weaknesses, opportunities and threats in accordance with known or existing boundary definition criteria. To this end, do weaknesses arise out of unknowns that may exist?
More importantly perhaps is the potential to identify these same unknowns and to establish ways of expanding the knowledge base to remove any uncertainties associated with them. For example the current political agenda associated with greenhouse gases is to reduce them. Can IVS determine the percentage global impact their efforts would create?
It may be prudent to retain this data together with the audit analysis as a separate appendix to the *business plan* owing to its potential to contain a large amount of information. Figures 22 and 23 discussed above should be used as an alternative approach for use with *business plan presentation needs.*
Figure 22 is used to define our SWOT analysis in presentation terms. It is suggested that for this example one matrix is used for the Industrial market and a separate one is used for the domestic market.

Fig 22 - Strategic Analysis – How are we going to get there?

The Marketing Plan Planning Process"

"Marketing SWOT Analysis – Industrial Segment"

	Internal	External
Good Points	Strengths ▪ IVS has the most suitable product mix across the market sectors ▪ IVS is noted for its quality, reliability and flexibility.	Opportunities ▪ IVS technology usage will adapt to market audit expansion needs. ▪ Revised health and safety requirements. ▪ Reduced energy applications.
Danger Points	Weaknesses ▪ IVS have become a reactive organisation. ▪ Market feedback impetus has lapsed.	Threats ▪ IVS management is not structured to accommodate expansion. ▪ Boundary definition criteria unknown for opportunities variables

Page 1 of 2

Fig 23 - Strategic Analysis – How are we going to get there?

The Marketing Plan Planning Process"

"Marketing SWOT Analysis – Domestic Segment"

	Internal	External
Good Points	Strengths ▪ Brand value.	Opportunities ▪ Market expansion arising out of global warming. ▪ New safety laws for the disposal of refrigerant. ▪ Revised insurance assessment requirements.
Danger Points	Weaknesses ▪ Market follower because of price. ▪ Sales orientation is allowing competitor image to improve.	Threats ▪ Boundary definition criteria unknown for opportunities variables. ▪ Exposure to overseas manufacturing risks. ▪ Short PLC/TLC

Page 2 of 2

We can examine how our SWOT analysis will or could impact upon our *business plan* strategy. This we will do using Figure 24.

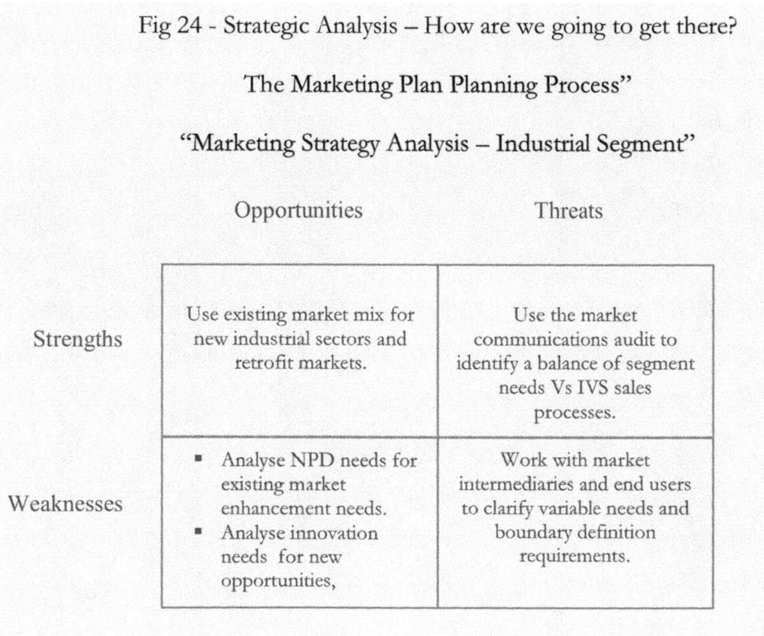

Fig 24 - Strategic Analysis – How are we going to get there?

The Marketing Plan Planning Process"

"Marketing Strategy Analysis – Industrial Segment"

	Opportunities	Threats
Strengths	Use existing market mix for new industrial sectors and retrofit markets.	Use the market communications audit to identify a balance of segment needs Vs IVS sales processes.
Weaknesses	▪ Analyse NPD needs for existing market enhancement needs. ▪ Analyse innovation needs for new opportunities,	Work with market intermediaries and end users to clarify variable needs and boundary definition requirements.

Page 1 of 2

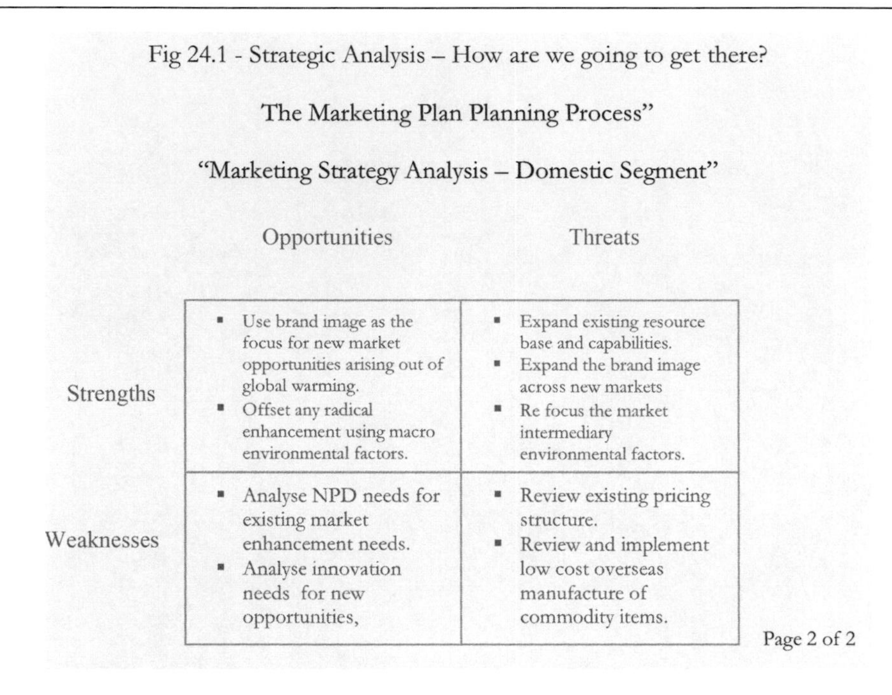
Fig 24.1 - Strategic Analysis – How are we going to get there?

The Marketing Plan Planning Process"

"Marketing Strategy Analysis – Domestic Segment"

	Opportunities	Threats
Strengths	▪ Use brand image as the focus for new market opportunities arising out of global warming. ▪ Offset any radical enhancement using macro environmental factors.	▪ Expand existing resource base and capabilities. ▪ Expand the brand image across new markets ▪ Re focus the market intermediary environmental factors.
Weaknesses	▪ Analyse NPD needs for existing market enhancement needs. ▪ Analyse innovation needs for new opportunities,	▪ Review existing pricing structure. ▪ Review and implement low cost overseas manufacture of commodity items.

Page 2 of 2

It will be noted that the latter two sheets have a common factor these being the need for a review of new product development (NPD) and to observe the impending need for innovation. Given that IVS are potentially embarking upon a radical growth and the fact that they are already brand leaders, competitors will eventually follow their lead. It follows that some form or precedence or bench marks be set for the industry macro environment in respect of the nomenclature employed with NPD and innovation. Nomenclature it could be said could stay with IVS for decades to come. For the purpose of this example we will use text book definitions to characterise our nomenclature.

- "Invention is a new technology or product that may or may not deliver benefits to customers.
- Innovation is an idea, service, product or technology that has been developed and marketed to customers who perceive it as novel or new. It is a process of identifying, creating and delivering new-product or service values that did not exist before in the market place.
- New-product development is the development of original products, product improvements, product modifications and new brands through the firms own R&D efforts."

Source: Kotler el al (2002)

One of the benefits of providing this particular type of information is the retrospective impact it may have on the strategies employed by IVS or indeed any other company pursuing this particular agenda. To this end, can we classify some of the existing product mix using the definition provided? This fact notwithstanding the nomenclature does have a level of importance in our analysis of its impact upon the *business plan* SWOT analysis opportunities will present. Figure 21 is used to identify in more detail, where IVS should focus their marketing strategies as further specified within Table 6.5.

Strategic Analysis & Steps 1 & 2	Step 3	Step 4
Industrial segment:		
A - Revised tendering processes	8	6
B - Revised contract handling procedures	8	6
C - Turnkey project activity	9	2
D - 3 extra heating outputs (Products A, B & C	9	2
E - 2 extra ventilation outputs (Products A, B & C	9	3
F - 3 extra ventilation outputs (Products A, B & C)	7	3
G - Wireless control (Products A, B & C)	6	7
H - Noise reduction solutions (Products A, B & C)	7	8
I – Maintenance	8	3
J - After sales	8	3
K - Retrofit applications	6	9
L - Re use of refrigerant	6	9
Domestic Segment:		
M - New products to meet global warming needs (Urban)	7	8
N - New products to meet global warming needs (Sub Urban)	6	8
O - After sales	7	3
P - Retrofit applications (Urban)	7	8
P - Retrofit applications (Sub urban)	7	3
Q - Re use of refrigerant	7	9

Table 6.5

Before plotting this analysis on our matrix it is to be noted that the values indicated for step 4 represent the impact upon the business in terms of competence availability. Perhaps unknown to text book authors it is possible to include additional columns and to produce additional matrices reflecting strategic analysis for individual business units and/or departmental activity. Here one could consider the use of this same matrix as a base to define additional or less resource base, office space, manufacturing space and so forth on a departmental basis. For this example the step 4 factors given above are an "across the board" average as determined by IVS.

It is also to be noted that certain of the factors are known to be required by the market audit but have not been rated highly. To this end, all of the development resource required to produce them cannot be undertaken concurrently. There exists therefore the need to analyse these in terms of operational and long term strategic marketing objectives and strategies as discussed within the next chapter.
Owing to the importance of this matrix, that is, strategic priorities it is suggested that they form part of the *business plan* presentation material. Depending upon the size of a given target market or market sector and strategic importance readers may wish to produce separate matrices for that particular segment or sector. For this example only two are used; one for the industrial segment and one for the domestic segment.
Our final comment associated with this chapter example consists of two specific criteria to be borne in mind.
In the first instance the "probability of occurrence" vertical axis employed above can be amended to suit the type of audit undertaken. This method of description (text book) is more associated with a market audit to reflect the market response to new or impending technologies (technology push). For this example the probability is high as the criteria provided are based upon market needs (market pull); we could in fact rename the vertical axis thus.

Text book authors would suggest that the marketing plan concentrates on the top left box forming part of figure 21. However, experience would show, particularly for the example provided that the top right box should also be considered Why, one would ask? The answer is that "quick hits" in terms of improving the margin can be achieved with minimal change to the structure of a company. Put another way, it provides the necessary kudos for individuals aspiring to be known within an organisation.

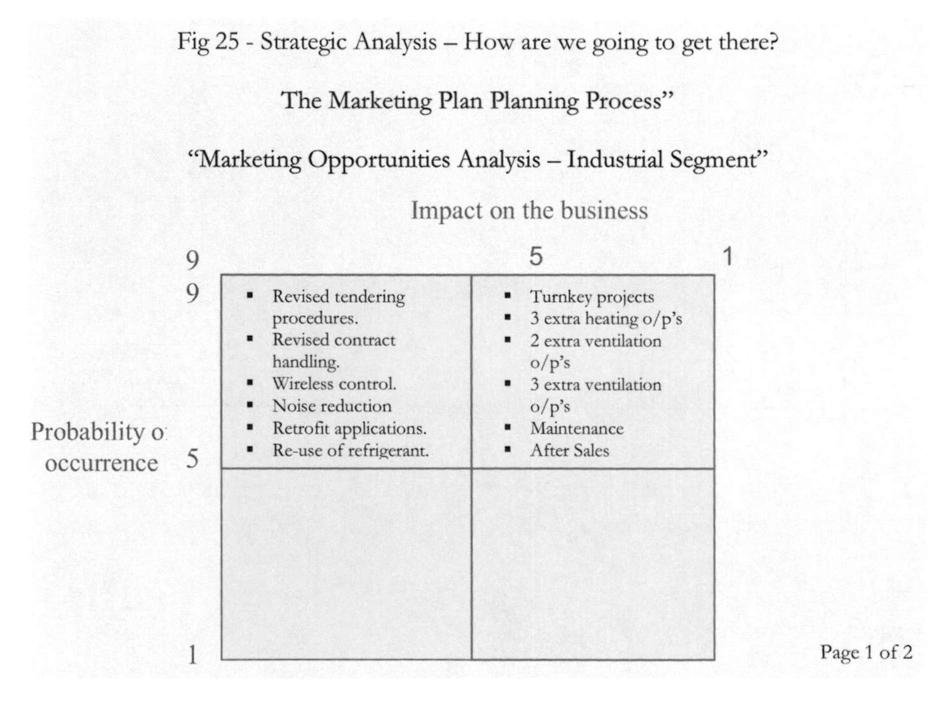

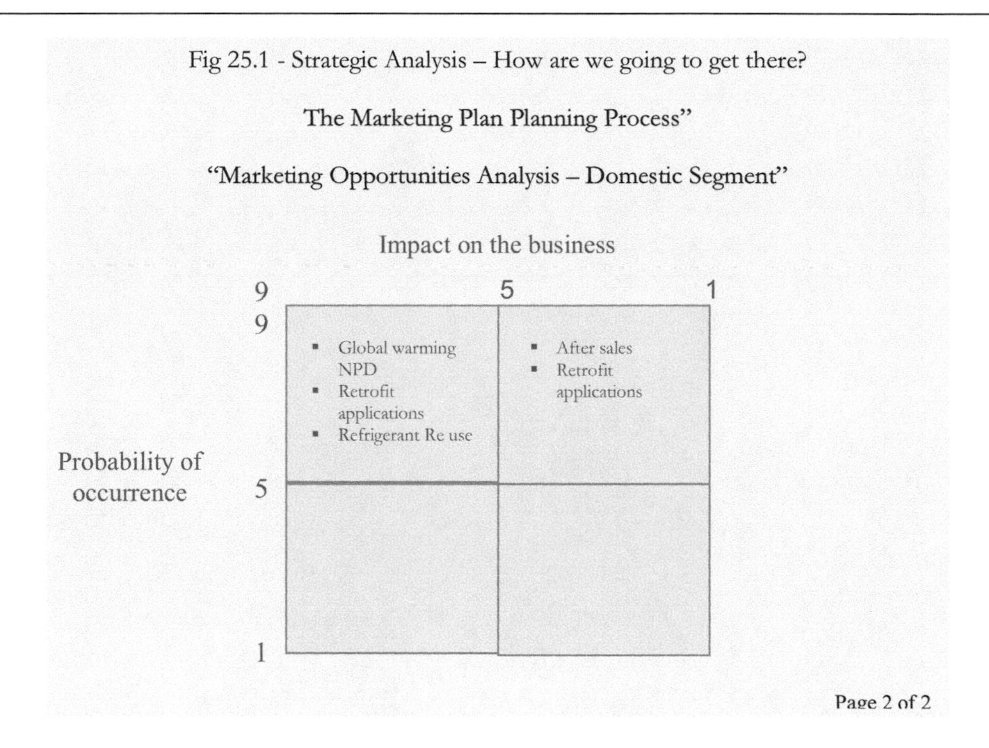

The remaining portion of this chapter is used to provide supporting text for environmental factor analysis. This supporting text will prove useful for our chapter 7.0 strategic planning exercise.

Supporting business plan research criteria

The macro environment we are told can be expressed in the form of PEST or indeed STEEPLE factors as already observed. Research has been undertaken where the macro environment is shown to have a direct bearing or otherwise on the research undertaken and its impact upon technology based market segments
It was observed using research journal data that additional factors can impact upon the PEST/STEEPLE factors. These are:

- Technological impact upon the environment
- Sales Impact upon the environment
- Logistical and purchasing impact upon the environment

Technological market segments have the potential to be populated with engineering and technology trained individuals. Whether these same individuals are unsure of their needs is unknown (Benkenstein and Bloch (1994, P. 15)). Their reaction to the deletion of products and the introduction of new products is similarly unknown (Hart (1989, P.9) and Wang and Montaguti (2002, P.83)).

Technology push principles employed within certain market segments would suggest not (research by Erfurt 2002). These same individuals may be academically trained but may not understand or know the best use of the technology (Cahill and Warshawsky (1993, P.17)). Technology push may also take on an assumption that the certain market segments by way of their culture are well informed of the technology being used (Slowikowski and Jarratt (1997, P.97)).

Ongoing reorganisation of the certain industries particularly in terms of change of ownership and redundancies could impact upon the sociological environment (Christopher (1970, P.80)). It is not known if this impacts upon the behavioural patterns within the industry. Cultural impacts could dictate the global point of origin of manufacture but the impacts upon the segment are unknown (Benkenstein and Bloch (1994, P.16)), Table 6.6 refers.

Author/s	Technological Impact on the environment
Benkenstein and Bloch (1994, P. 15)	Suggest that technological uncertainty creates unstable demand patterns with those consumers who are unsure about their real preferences or the benefits they will get from the technology.
Benkenstein and Bloch (1994, P. 16).	Expands upon Christopher (1970, P.80), they suggest cultural changes may be important when considering the geographic technological market boundaries that exist on a global basis. Product and alignment and the technology it uses with the market it is targeting may also have an impact upon the problem solving aspect of customer needs.
Brooksbank et al (1999, P.117).	Consider market needs and the importance associated with responding to them but do not refer to the environment in their research.
Cahill and Warshawsky (1993, P.17).	We are warned of the potential to presume that there exists “technological people”
Christopher (1970, P.80)	Expands upon Slowikowski and Jarratt (1997, P.97) by suggesting that changes in the cultural environment mirror the changes that are occurring with the sociological environment. He suggests that consumer behavioural patterns are changing in terms of attitudes and perceptions towards products.
Crick and Jones (1999).	Add another potential environmental factor - ergonomics
Hart (1989, P.9)	Argues the importance of considering the environmental

	impact upon the companies researched on product deletion strategies. He concludes that environmental factors may impact upon both the long term and short-term strategies. He further concludes the importance given to strategic marketing issues implies a lessening of pressures associated with product deletion strategies.
Slowikowski and Jarratt (1997, P.97)	There is evidence to show that marketing managers in industrial segments assume that the client base is well informed of product technology. They argue that where culture may dictate different marketing strategies they should be employed in a manner that focuses upon the culture they are aimed at
Wang and Montaguti (2002, P.83).	There would also seem to be an area of reaction within the macro environment that prevents or detracts upon new product introduction into the market. They state that very little research has been undertaken within this field of marketing.

Table 6.6

Sales Impact on the Environment

Any SWOT analysis in respect of "Sales" impact will relate more to the processes employed by the company. A sale is defined as the exchange of money for goods (Collins dictionary (1996, P439)). Evidence exists (Erfurt 2002) that sales processes employ sales individuals only in part of the sales process (Jobber and Lancaster (2003, P.121) and Tansu Barker (2001, P.22)). This suggests that the definition of sales individuals should be changed either by way of job title and/or work definition. The postulation of re-definition put forward for this example is "order takers" for those sales individuals who are not active within the contract process through to client payment, Table 6.7 refers.

Author/s	Sales Impact on the Environment
Jobber and Lancaster (2003, P.121)	Stress the importance of analysis arising out of the sales process problem solving and need analysis. It should occur early within the sales process.
Moncrief and Cravens (1999)	Failed to recognise Jobber and Lancaster (2003, P.121) observations when reviewing the importance of selling and sales management within their research.
Tansu Barker (2001, P.22)	Considers the competitive, the economy and selling and

	measurement of sales organisation environments.

Table 6.7 Sales Impact on the Environment

Logistical and Purchasing Impact on the Environment

Owing to the potential paradox stated in terms of "sales" and "order takers" the logistical impact on the sales of mature products may not it is suggested be gauged effectively for certain market segment (Darden et al (1989, P.48)).
Purchasing barriers may exist within segments but this it is suggested is dependent upon the type of product it applies to within a market segment comprising of many products and any technology push techniques that are employed with these products (Czuchry and Yasin (1999, P. 240)), Table 6.8 refers.

Author/s	Logistical and Purchasing Impact on the Environment
Czuchry and Yasin (1999, P. 240)	Suggest that it is the purchasing process that can cause barriers at the outset. They may not be as receptive to technical innovations as other individuals in the same organisations.
Darden et al (1989, P.48)	Argue the need to consider logistics within the marketing process. They suggest that observing "customer needs" will have a substantial impact on the sales of mature products

Table 6.8 Logistical and Purchasing Impact on the Environment

This section is seen to provides us with a number of additional variables for consideration, namely the ergonomics, logistical, purchasing variables as well as the sales impact upon the macro environment if not the technological environment. There is also the turbulent impact that technology can impact on the environment as given within Table 6.9.

Author/s	Comment	References
Dhanani et al (1997, P.160)	Rapidly changing nature of high tech products and their markets, which result from such factors as frequent product improvement, major technological breakthroughs"	Davidow, W.H. *Marketing High Technology,* The Free Press, New York, NY 1986 (Note this is out of print)
Dunn et al (1999, P.186)	High tech firms operate in uncertain and volatile environments where technologies,	Bahrami, H and Evans, S (1989), "Strategy making in

	technological boundaries and market conditions change continuously"	high-technology firms: the empirical mode", *California Management Review* Vol. 31 No. 2 pp 107-28
Benkenstein and Bloch (1994, P.15)	Technological uncertainties give rise to a lack of clear definitions, few barriers to entry and significant sized economies.	Meldrum, M.J and Millman, A.F (1991). "Ten risks in Marketing High-technology Products", *Industrial Marketing Management*, Vol 20, pp 43-50.
Beard and Easingwood (1992, P.5)	"Factors contributing to this uncertainty include short product life cycles (PLCs), frequently changing market conditions and industry boundaries which are blurred by frequently changing market segment conditions"	Capon, N and Glazer, R "Marketing and Technology" (1997): A Strategic Coalignment", *Journal of Marketing*. Vol. 51 No. 3, pp.1-14

Table 6.9

In terms of journal resource data it is to be noted that there is a lack of reference to the macro environment which is difficult to understand. Journal authors are more concerned with marketing research within a microenvironment but appear not to follow an "option identification", "forecast outcomes" and "evaluate outcomes" research process (Wensley (1989, P.70)) or similar method of supporting criteria for their research.

Trustrum (1989, P.48) argues that companies who do not adopt an appreciation of the macro environment become concerned only with its own capabilities and values. He goes on to suggest that the task of a marketing based organisation is to match organisational capabilities to the demands of a particular market segment.

This author would suggest that there is the potential to research the complexities of technological environmental factors. Scientific evidence is available to show that technologies can be observed, monitored and predicted (Palmer and Williams (2000)). They conclude;

"There is an apparent correlation between development activity and rate of technological progress"

Formulate Strategies

Substantial data is available that will assist with the preparation of business (marketing) strategies and assumptions. This same data, it will be seen, can be used to produce an effective ***business plan.*** What is important however is to bear in mind that strategies are the means by which marketing objectives will be achieved. Strategies will involve a process of planning but before this planning can take place some form of assumptions must take place in relation to key determination success factors.
One text book author stresses the importance of strategy formulation as follows:-

> "Those who do not study the past will repeat its errors.
> Those who do study the past will find other ways to foul up"

Source: Piercy (2003, p. 375) *Market-Led Strategic Change "A guide to Transforming the Process of Going to Market. (third edition).*

In reality we need to be concerned with what objectives and strategies are important to the business. Text book criteria would suggest two statements for consideration (Piercy (2000. p. 570).

- "*Intended Strategy or Strategic Intent* – what we think or what we want the business to be about in the marketplace. Typical factors that can impact upon this decision process are given in Table 7A:-

Strategy	Required Effect
Do Nothing	You wish for a peaceful co-existence.
Organisational Functionality	"Sales" orientated? "Marketing" Orientated?
Direct Competition	Be price conscious (Sales orientated)
Indirect Competition	Product Differentiation Promotional Differentiation Distribution Differentiation } Market Orientated
Innovation/Invention	Change the rules of the game
Withdrawal	Market exit

Table 7A Modified basic strategies open to a company (Baker 2000)

- *Perceived Strategy or Strategic Reality* – what the business is *actually* about in the marketplace, as it is perceived by the business, and ultimately as it is perceived by the target customers (the ones we have *lost*, as well as the one we have gained, Incidentally)"

The same author summarises the statement by alluding to these statements as the objectives and strategies associated with "Getting the act together". The analogy of the latter is given as:

> "One of the major practical problems we face in dealing with this lengthy, overlapping, conflicting, uncertain, messy and complicated set of issues which make up market strategies, marketing programmes and plans is packing the whole thing together"
>
> Source: Piercy (2000. p. 570) *Market-Led Strategic Change "A guide to Transforming the Process of Going to Market. (third edition).*

Prudence would suggest that no science should concentrate on providing a formal strategy for all of the individual business activities that inhabit the planet. Clearly, many variables exist within all market types each with a varying degree of boundary definition. The marketing audit and SWOT analysis, to some extent should have alerted business owners and managers of their existence and the potential opportunities that exist for their own market activity. Regrettably however, there will never be sufficient statistics that will enable a truly balanced approach to formulating strategies and as such assumptions will need to be made to create a means to an end.
The latter fact notwithstanding the use of assumptions can be made more realistic and accurate. Their use impacts more upon forecasting which is discussed within the next chapter and as such this chapter looks at how to build a market offering in relation to the previous chapters.
So which is the best approach to take?

In the first instance any strategies employed will be determined by the marketing objectives. It is the SWOT process in respect of the marketing audit that defines the marketing objectives. Similarly, these marketing objectives must align closely with *business plan* objectives; indeed, most probably steer them. These may refer only to products and markets in that, strategies are mainly concerned with the four P's, of the marketing mix as shown in table 7B.

Product:	The general policies for product deletions, modifications, additions, design, branding, positioning, packaging etc.
Price:	The general pricing policies to be followed for product groups in market segments.
Place:	The general policies for channels and customer service levels.
Promotion:	The general policies for communicating with customers under the relevant headings, such as advertising, sales force, sales promotion, public relations, exhibitions, direct mail etc.

Table 7B Adapted from McDonald, M. (2003). Marketing Plans: *How to Prepare Them, How to Use Them (fifth Edition).*

Here the marketing plan is used to analyse the current business scenario as well as that planned for the future. To this end, all business owners and managers should take a close look at the products or services offered. Marketers use the expression "portfolio management" to describe this activity. Text book data provides us with two specific models for use as directional aids to portfolio management assessment as given in Fig 26 & 27.
Fig 26 should be used to assess the individual services that you provide be it a basic product or superior product or additional services. Each one can then be considered in respect of each other in terms of the potential for growth for that activity and hence, any investment allocation.

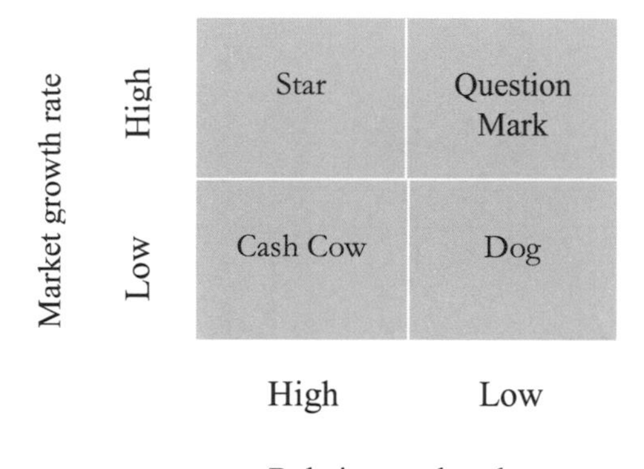

Fig 26 Directional aid to portfolio management "BCG Growth Matrix"

Clearly, a bank manager or other interested parties associated with a loan may have an interest in this model as well as that given for Fig 27 in particular the process used to arrive at the production of these models. Included within Fig 27 are four circles indicated by a letter. The four letters used represent four products or service activity of any given business. It could be that only one or two such activities exist or perhaps even more.

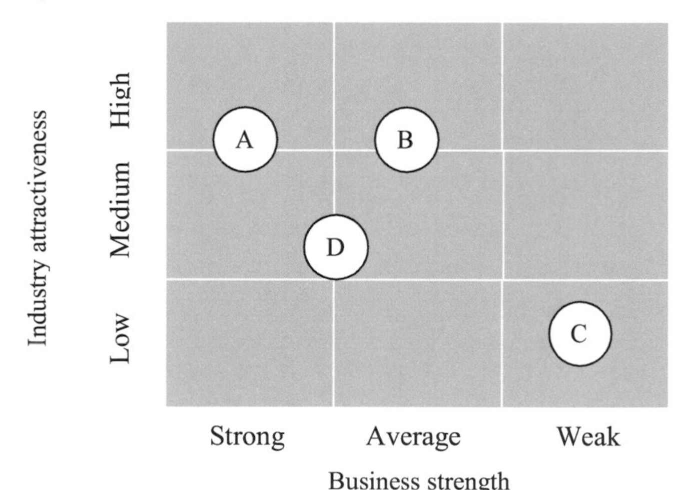

Fig 27 **Directional aid to portfolio management** "GE Strategic business planning grid"

Irrespectively, there is a need to understand the relative positions of each service or product activity. Circle A could be one specific item of the product mix the business offers of which the market share is high within a reasonably sized or large attractive market. This indicates a healthy scenario worthy of any further investment to sustain the service offered.

Services B & D on the other hand show average to strong percentage share in the same and smaller markets respectively which would indicate a higher level of investment to boost market share and market expansion respectively. The investment considered should it is suggested be used to change or improve the service in order to make it more attractive in that segment. Service C is a weak activity in a small market and thus would suggest its deletion from the business product/service portfolio or one that requires constant monitoring.

Constant monitoring it must be said will consider more the management of the relevant business unit. Imagine the monthly if not weekly reports on performance that will need to be produced for product C in comparison to that for product A.

It can be seen that Fig 27 is useful when assessing what direction the business should take and thus should impact upon the policy decisions undertaken by business owners and managers. Here it is important to remember the maxims associated with the 4Ps as given within Table 7B and their principal issues and association with business policy. Summarising their effect on the latter is given below.
Here again the data would need to be presented in bullet point format for any *business plan* presentation.

7.1 Product Policy (adapted from N. Piercy 2003)

Product Policy	Intended Effect
Product definition	Consider the services offered as a single entity as well as a grouped package in relation to competitor products. Consider also the quality of the product, the design of the product as well as any specific or unique product features.
Product mix selection	How do these serve the target markets in a way that makes sense to prospective customers as well as creating value?
Branding policy	What message will you use to convey to prospective customers your competitive positioning in respect of the value that you are providing?
New product assessment	Always bear in mind emerging customer needs as well as the need to fill any gaps in the existing product range or to replace obsolete products or services

Outsourcing	What are your core competences that create difficulties for your competitors? Separate bespoke competences from commodity based competences. Consider outsourcing activities that could be sourced less expensively by others or by manufacturing in other regions.
Managing product or service deletions	This will always be a major issue for many business owners particularly with parts availability. Here there may be a need to revert back to the SWOT analysis.
Target markets	The marketing audit will most certainly give rise to a mix of products requirements to an extent that it is unlikely that the variables given hereof will create the same intent for all target markets. The ability to adapt more easily than competitors must rank high when undertaking the SWOT analysis.

Table 7.1

7.2 Pricing Policy (adapted from N. Piercy 2003)

Pricing Policy	Intended Effect
Price positioning	Consider the price of your product against competitors but perhaps more importantly in terms of what customer expectations are. Consider also any low prices in terms of volume requirements but review the potential to escalate any particular product benefit or service not offered by others. Exchange rate variation will also impact upon the positioning techniques employed as discussed in the "Pricing Audit" – chapter 5.0 and its business impact discussed immediately below.
Price levels and business impact	Review Fig 10 above in respect of the margins each product or service offers for any change in strategic intent. Reductions in margin as well as any increase in margin associated with exchange rate variation are likely to be of a short duration. This suggests that one particular product could be rated as a cash cow for one or two months perhaps.
Discount structure	Each market segment has the potential to attract a different discount structure as the packages offered for each market segment will vary. Namely, the ones where product or service may vary or has the potential to vary in respect of the package offered.

	Conversely, if considered a luxury, quality and reliable brand do you want to offer any discounts? It could be that potential clients are looking for your type of product and are prepared to pay a premium (SWOT analysis?)
Market intermediaries	Sales channels are used in most business activities. Here one must understand fully the macro environment within which the company operates. Micro environmental (within the macro environment) needs must always be assessed in relation to end user needs and the sales channels adopted to meet such needs, see Fig 40 page 221.

Table 7.2

7.3 Place Policy (adapted from Kotler et al; McDonald)

Place Policy	Intended Effect
Product availability	What activities do you undertake to make the product available to all of your target markets? These activities, in the main may rely upon direct marketing channels and channels of activity undertaken by others. However, in what way do the others undertake selling on your behalf? The main criteria should be in what way you can support their selling activities in terms of availability as well as ensuring availability (for the target markets) that aligns with the overall marketing concept. Yes the internet is often viewed as a panacea in respect of the latter but will, it should be noted apply more to consumables where the only requirements is to fulfil a basic need at the lowest price available.
Product delivery	Boundary definition criteria will determine the overall policy associated with this requirement. One such policy would be to review the product or service differentiating techniques employed.
Supply logistics	What logistics are in force to ensure you can support the sales efforts by others in terms of that required to meet operational and long-term strategic needs? Similarly, what logistics are in force that will enable you to question the services offered by intermediaries in a way that helps you to ensure market needs and profit maximisation are more balanced.

Table 7.3

7.4 Promotions policy (adapted from N. Piercy 2003)

Promotion Policy	Intended Effect
General comment	It is argued that the advent and growth of the internet has had both a positive as well as a negative impact upon advertising policy. Here it is suggested that sellers must now adapt to the buyers purchase process. To some extent this could be as a result of the rise and dominance of marketing intermediaries that do not possess academic based marketing skills. However, do not forget the potential to directly market any niche services offered. Here, the potential will always exist for the buyers to adapt to the services offered.
Advertising	In the first instance it is important to note that internet based business refers only to those individuals who are seeking a product or service arising out of a need. Clearly, advertising is used to also create that need using mass media such a TV, radio, the press as well as billboard type methods. It is suggested that following the processes used to position and differentiate the business that an audit be undertaken of all the channels used already to advertise the business in terms of the positioning and differentiating techniques applicable to the *business plan* and those currently employed.
Personal Selling	Knocking on doors both locally and nationally (depending on target market) together with supporting identification data will create an interest that may not have been there already.
Sales promotion	What events are happening within your geographical region that has the potential to impact upon your business? The Olympic games for instance and the manner in which the events are distributed geographically create the potential for promotional review, such as incentives or price cuts for a given product offering. The emphasis here is any event that has the potential for the business to share the same advertising media as the major potential activity.
Public relations	Using "Fawlty Towers" as analogy Basil endeavours to alter his

	image depending upon the relative disposition of the guest, mostly unsuccessfully. Text book data suggests successful public relations takes into consideration your corporate image in relation to the different audiences you are targeting. This needs to be borne in mind if offering 5 diamond hotel accommodation to "Rab C Nesbitt" who has just won the lottery.

Table 7.4

Any assumptions that remain at the end of this chapter can be listed as a company's own assessment of itself and the assumptions made. Always remember that they can be a source of opportunities and threats. However, any impact the latter may have on policy can be offset by observing the competitors own marketing strengths, in particular their value chain.

7.5 Chapter Summary - Strategy Formulation Measurement Techniques

To all intents and purposes the strategy techniques put forward within this chapter are more related to marketing techniques and may not to a large extent encompass *business plan* needs. There is a need to summarise the outcome of the work undertaken in respect of its impact upon the projected business as a whole, in particular the P&L as reviewed within the next chapter. That is, in what way can we measure the impact that our *business plan* will have on the business?
We can show these efforts in the form of current activities compared to *business plan* suggestions or projections. These can take the form of that shown in Table 7.5:-

Marketing Mix	**Action**	**Outcome**
Product Policy	Increase target market base by increasing quality and reliability	Increased number of quality enquiries
Pricing Policy	Increase manufacturing efficiency to lower costs.	Maximise profitability
Place Policy	Purchase of new machinery to reduce delivery periods.	New differentiation factor.
Promotions Policy	Realign marketing expenditure in respect of target market trends	Increased customer satisfaction

Table 7.5

Experience would suggest that this information be produced on separate sheets if operating within a large organisation and undertaken by the relevant individual departmental managers.
It is to be noted that the measurement activities stated serve to analyse only several departments within a given organisation.

It is suggested that company strategy be reviewed on current performance in relation to planned/suggested performance improvements or definitive actions. It follows that any other departmental activity that exist within the company produces corresponding strategic data. More importantly, in what way will these other departments' support the strategic activities given as suggested in Table 7.5.1?

Business Support Activity	Action	Outcome
HR (Personnel Dept)	What training requirements will be needed for all of the employees associated with the business plan strategic needs?	Market competency needs support base is sufficient to meet the projected financial goals.
Finance Department	Re-evaluate financial software employed to evaluate operational financial performance	The ability to improve the response criteria associated with analysing and improving any and all costs that impact upon the efficiency of the marketing mix.
Legal Department	What legal barriers may exist that may prevent the implementation of strategic needs.	Faster process to market and improved sales process (state periods involved (e.g. bid response times)

Table 7.5.1

Remember that the submittal of any data in respect of the summary put forward must relate to strategic activity or intent and not departmental objectives. Keep in mind that these are the strategies you envisage to meet the corporate objectives.
Similarly the macro environment within which the company operates may serve to expand the summary provided and the strategies employed. Figure 17 chapter 5.0 would suggest that certain companies will trade direct with end users but also with market intermediaries. It follows that our strategy could be defined in relation to "Direct Sales" and/or "Indirect Sales" where the latter considers micro environmental trading activity (through a marketing intermediary). Here the

marketing intermediary would have responsibility for all promotional activity (and costs) for specific target markets.
Direct sales on the other hand would suggest that table 7.5 be expanded; table 7.5.2A refers:-

<table>
<tr><th colspan="3">Direct Sales</th></tr>
<tr><th>Marketing Mix</th><th>Action</th><th>Outcome</th></tr>
<tr><td>Product Policy</td><td>Increase target market base by increasing:-<ul><li>Quality and reliability.</li><li>Product Features.</li><li>Product Packaging.</li></ul>In relation to individual target market needs.</td><td>Increased number of quality enquiries</td></tr>
<tr><td>Pricing Policy</td><td><ul><li>Increase manufacturing efficiency to lower costs for all markets.</li><li>Set market price levels in relation to product policy compared to target market needs and competitor activity.</li></ul></td><td>Maximise order intake and profitability</td></tr>
<tr><td>Place Policy</td><td><ul><li>Purchase of new machinery to reduce delivery periods.</li><li>Logistics activity to be aligned with individual target market needs.</li></ul></td><td>New differentiation factor.</td></tr>
<tr><td>Promotions Policy</td><td>Realign marketing expenditure in respect of target market trends</td><td>Increased customer satisfaction.</td></tr>
</table>

Table 7.5.2A

Whereas Indirect sales would suggest that table 7.5 be expanded in a different manner; table 7.5.2B refers:-

Indirect Sales		
Marketing Mix	**Action**	**Outcome**
Product Policy	Increase target market base by increasing:- • Quality and reliability. • Product Features. • Generic product base In relation to individual needs.	Increased number of sales and manufacturing efficiency. Ability to create generic standards.
Pricing Policy	• Increase manufacturing efficiency to lower costs for all markets. • Set market price levels in relation to individual needs and competitor activity. • Negotiate market intermediary added value and costs.	Simplified and less costly sales process.
Place Policy	• Purchase of new machinery to reduce delivery periods. • Logistics activity to be aligned with individual target market needs.	Simplified accounting and billing procedures/processes
Promotions Policy	Realign marketing expenditure in respect of target market trends	Emphasis will be on supporting others.

Table 7.5.2B

Last but not least is the need to consider the integration of all the strategies employed in respect of the sales process. For the purpose of this book it is the departmental activity that occurs from the work required to produce an enquiry through to the receipt of 100% payment. It is the activity that is of paramount importance in respect of the overall financial efficiency of the business.

International standards are available (ISO9000) or its latest derivative that assist with the planning of sales processes but even these do not prevent the potential for fragmented activities to occur within the process. Nor may they consider chronological department overlap needs. Experience would show that these potential deficiencies occur within "Sales" based organisations rather than "Marketing" based companies. Put simply, the former is more likely to have a marketing mix that is not balanced with market needs which gives rise to more sales process activity. This can be represented as shown in Fig 28.

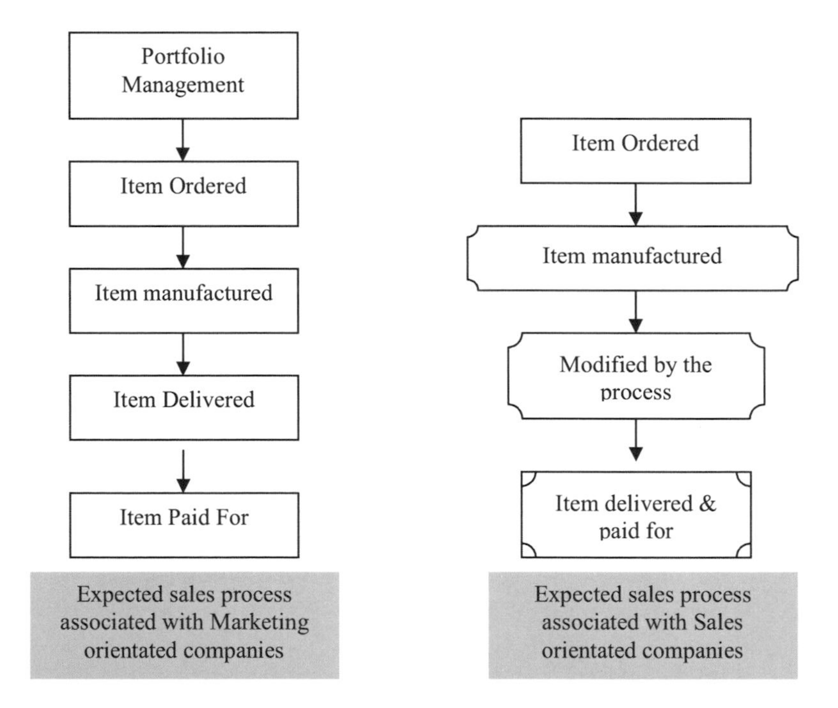

Fig 28 Chronological Sales Processes

Sales processes are those processes employed to balance the company structure in respect of market needs and thus form part of the marketing plan and *business plan* strategy formulation techniques. They will have an inherent association with maximising throughput efficiency in respect of a documentary alliance with marketing plan chronology but in a transparent manner.

For example those criteria that impact upon SWOT analyses and consequent strategy formulation must form an inherent part of any process. Another factor for consideration that may not be easily recognised as part of any audit or SWOT technique is the ability to undergo any contract change during the delivery period. Here there is a need to ensure that any such change does not create a major impact upon the sales processes employed. Its impact is to remove any "take it or leave it" philosophy which may be suitable for one off orders but not where one is seeking to create an impact upon a market segment.

The effectiveness of any sales process can be assessed in terms of the documentation used as part of our implementation programme as discussed further within chapter 9.0 hereof. However as a matter of importance sales process effectiveness and levels of efficiency will be determined by our strategy formulation in respect of the preceding aspects of our marketing plan. Documentation must be provided that states as a standard what our portfolio management will comprise of as a standard.

This will have the effect of informing the client base as well as any vendor what the variables are that form part or whole of any given contract and the boundaries associated with them. Still found in the USA this method of trading can be observed as part of the promotional documentation provided by many vendors. Here they will state what they do, the options available and the limits of use with all aspects of the product under consideration. Some vendors will actually describe and compare/tabulate product or service benefits with competing companies.

Similarly techniques are employed by large Japanese and Italian based contractors to improve and simplify as well as reducing the cost of their enquiry processes. To this end, bespoke documentation is not used for individual contracts. Rather the engineer and purchasing officer will request a specific specification/commercial or legal document from their documentation department and then will annotate (tick list) accordingly for the subject at hand. Here a generic document remains constant but is amended to suit specific project needs.

7.6 Chapter example

Having completed half of the business plan half of the *business plan* there is a need to be cognisant of the fact that all the work undertaken this far aligns closely with corporate strategic needs. The completion of the SWOT analysis may require further discussion with the management team to establish the strategic intent of the business. This is given in presentation format, fig 29 refers:

Fig 29
Strategic Analysis – How are we going to get there?

The Marketing Plan Planning Process"

"Marketing Objectives – Strategy Formulation"

Page 1 of 3

Fig 29.1
Strategic Analysis – How are we going to get there?

The Marketing Plan Planning Process"

"Existing Strategies"

Organisational/Operational Functionality

- Initially market needs based but we have evolved into a "sales" based organisation.

Competitive Nature

- Technology conscious for specific industrial market segments.
- Reliant on brand image for the domestic market segment.

Page 2 of 3

Fig 29.2
Strategic Analysis – How are we going to get there?

The Marketing Plan Planning Process"

"Strategy Formulation – *Business Plan* Strategic Intent"

Organisational/Operational Functionality

- To become a marketing orientated organisation operative in a sales based macro environment.
- To maximise our opportunities and minimise our risks arising out of operational and long term objectives.

Competitive Nature

- To be price and technology conscious for all market segments.
- To differentiate IVS for all market segments through product and distribution using NPD, innovation and invention.
- To withdraw from any market segment where price is the only known market variable.

Page 3 of 3

The latter bullet point will have significant relevance on the *business plan* as will be outlined further within this chapter example.

We must now examine our product portfolio by way of our existing mix and its ability to meet operational needs as well as what is required to maximise the available potential as identified within the market audit.

For the purpose of *business plan* documentation presentation needs a BCG matrix (Fig. 26) must be produced for each of the products for the existing market segments. For simplification purposes they are not shown here mainly because they do not encompass our main aim of maximising our opportunities. Clearly, the additional market sectors observed and the existing sectors all have growth potential but in the main they are reliant upon an expansion of the existing product mix.

It is for this reason that our example will concentrate on examining the GE strategic business grid – Fig. 27. For those of you who may argue that the vertical axis relates only to industry attractiveness let us not forget that the marketing plan process employed this far has already identified each market sector in terms of monetary value. It follows that the industry under scrutiny is already considered a growth market.
Here again we will limit the number of sheets for this application by producing only two examples, fig 30 refers. When assessing and analysing potential new market segments or sectors where sales do not currently exist one should use the market audit to gauge or anticipate positioning based on the known competitor analysis derived from the market audit.

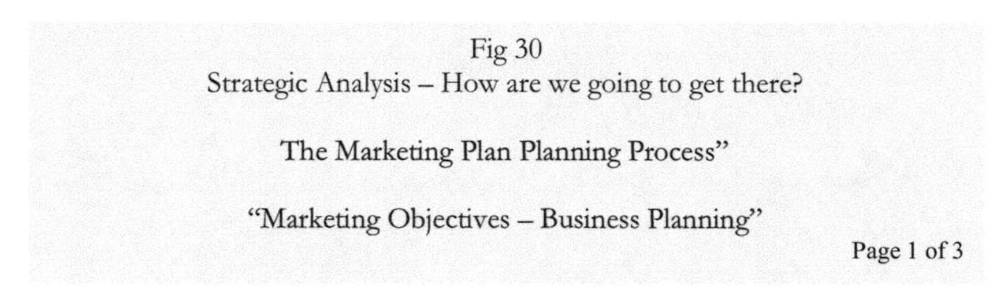

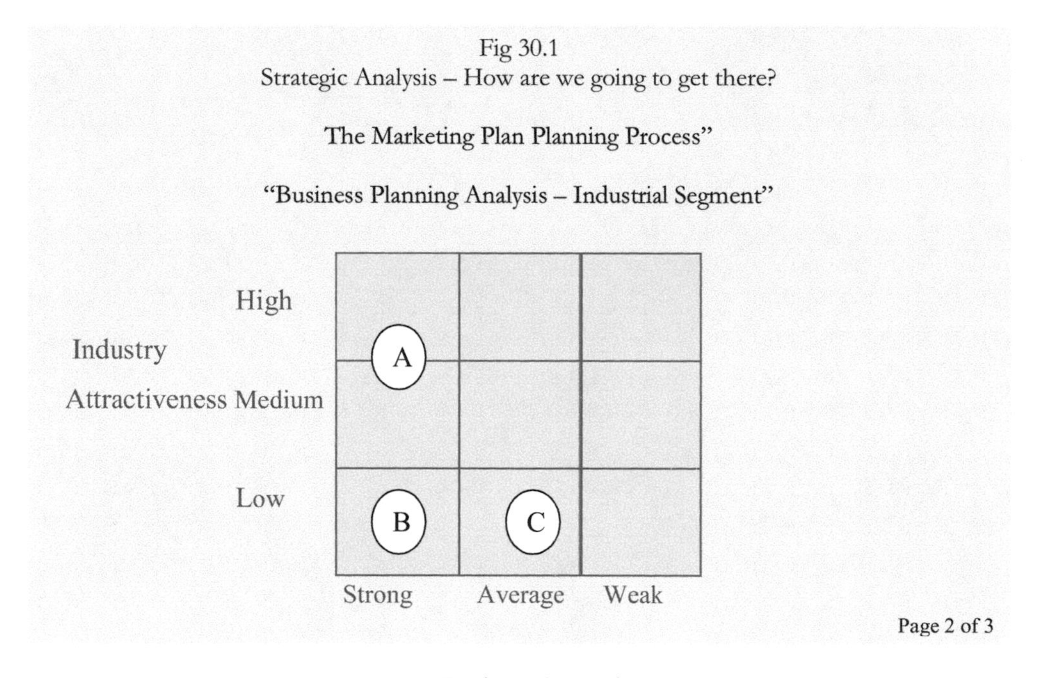

Business Strength

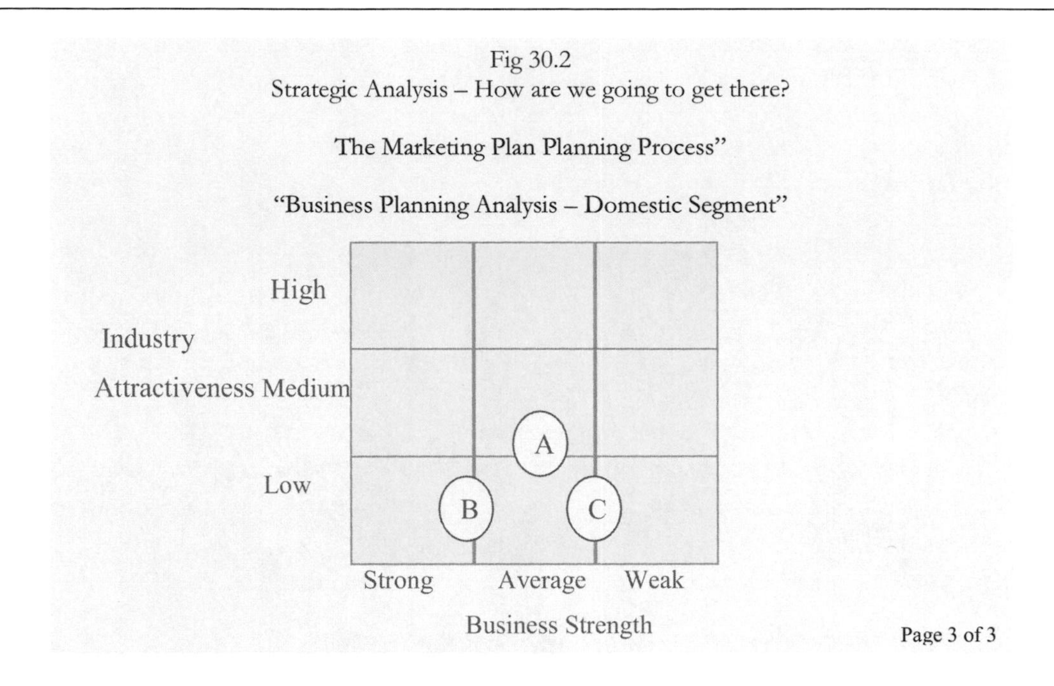

Next begins our analysis of what factors have determined the strength of our portfolio within the market segments shown. More importantly perhaps is what is it we must undertake to not only improve our standing but to define our product place, price and promotional strategy as a whole (Directional Policy).
At this time there is a need to produce for each of the departmental managers a list of variables and boundary definition criterion (associated with the market audit) that they must respond to in order to provide a formal directional policy for the overall *business plan*.

Product Policy
In what way will the product policy increase industry attractiveness using existing competences? In what way will the product policy increase industry attractiveness by increasing the existing competence base?
Note that the wording used (and for the remaining policy charts) takes into consideration that the questions raised were implemented within the market audit undertaken within chapter 5.0. The need to undertake an iterative review of a marketing plan is stated within Table 7.6

Product Policy	Actions To Be Undertaken	Strategic Impact
Product definition	Industrial segment ▪ Redefine basic design criteria for all products. ▪ Establish and separate commodity based design (common to	Industrial segment ▪ Basic materials will remain but packaging will change. ▪ Basic materials are a commodity item, integral design and

	competitors) components. What is it that enhances our quality & reliability?	assembly and test procedures enhance quality & reliability.
	▪ Establish and separate unique IVS features. What is it that enhances our quality & reliability?	▪ Technology associated with the electrical system employed and manufacturing processes are the factors that enhance quality & reliability.
	▪ Technology can cause market turbulence in what way can we remove this deficiency?	▪ New technology is under review to replace existing control system.
	▪ What differentiation factors can we attribute to our maintenance services offering?	▪ New technology above will monitor life of components and create wireless access for maintenance staff.
	▪ How must we package our refrigerant re use product for macro environmental usage?	▪ New pump is under development patent pending.
	▪ In what way does product design simplify and maximise after sales service opportunities?	▪ New technology integrates more components into one package – less parts to be used.
	Domestic segment (generally in accordance with above)	Domestic segment (generally in accordance with above)
	▪ Redefine basic design criteria for all products.	▪ Products A & B not considered suitable for long term strategic needs. Product C will be re-designed.
	▪ Which technology must be employed to accommodate the impact of global warming?	▪ New product D under consideration for use with solar heating systems. a) New product E under consideration using micro wave technology to reduce cost of

	▪ Which components are suitable for overseas manufacture?	domestic heating. b) New product F under consideration to integrate domestic refrigerant usage ▪ Product C (low cost sub urban domestic option) in its entirety.
Product mix selection	Industrial segment ▪ Do we use the same product design for all industrial sectors? ▪ In what way do we accommodate the different options associated with each separate sector? Domestic segment ▪ Hotels are now classified as low end industrial plus domestic – how will our mix selection vary across the sector? ▪ What technology features/benefits must be employed to maximise early adopter (Hotels, Office buildings & Urban) sales arising out of global warming. ▪ In what way can our competence increase market attractiveness of all domestic sectors?	Industrial segment ▪ The commodity items are standard but are designed, manufactured and delivered in bespoke format. ▪ Individual needs are determined by rated output which in turn requires different software options. Domestic segment ▪ Product C will be modified to incorporate a range of outputs that align with the audit. ▪ The control system will undergo significant change to make it less complex. Some of the industrial benefits will be adopted for the office building market sector. ▪ Our technology and competency usage should be assessed in terms of brand image rather that overall domestic market segment share.
Branding policy	Brand image already exists but how will our product policy enhance our brand image?	The corporate strategy associated with our marketing plan will enhance our technological

		leadership, quality and reliability standing across the industrial and domestic macro environment.
New product assessment	Industrial & Domestic Segments ▪ To analyse NPD and innovation/invention using this product policy in respect of our corporate strategy and marketing process employed thus far. ▪ What costs and timescales are involved?	Industrial & Domestic Segments ▪ Variable and boundary definition criterion employed previously for the new developments stated above have been amended to reflect marketing plan needs. ▪ These details are produced in a separate presentation sheet as given below.
Outsourcing	Industrial & Domestic Segments ▪ Identify the parts of our sales process that has the greatest impact upon user quality and reliability. ▪ What competence is used to create quality and reliability?	Industrial & Domestic Segments ▪ Compatibility and completion of User needs Vs IVS supply standard documentation. a) Inter departmental information interchange. b) Non intrusive document change system. c) Manufacturing machinery. d) Efficient costing system e) QA & testing procedures. f) Expanded installation procedures ▪ Standard but easily adapted tender/contract documentation. a) Overlapping sales process documentation. b) Manufacturing work force skills base. c) Contract management skills. d) Management skills base

	▪ Compare actual costs with estimated costs.	e) Installation skills work force base ▪ Calculated every month using a computer based system. Details known only to Mohamed, Manfred and Eugene.
Managing product or service deletions	Industrial & Domestic Segment ▪ If products are deleted how long will spare parts be available? ▪ Is deletion policy caused by inability to meet health and safety and/or legal requirements? ▪ Is the replacement fully compatible with old? ▪ Do we have to undertake a re-training programme?	Industrial & Domestic Segment ▪ Not previously discussed but market audit suggests a minimum of 5 years. ▪ Product B will not meet with anticipated new H&S office building regulations (as per audit). External legal assistance has been sought to identify type of wording used to notify market. Company insurance is also under review. ▪ Not previously considered but will now introduce a programme of implementation which will enhance the retrofit market opportunities. ▪ Only for the industrial market segment. New installation manuals will be produced for the domestic market segment.
Target markets	Industrial & Domestic Segments ▪ Do we need to redefine product specifications for each market sector and the options available?	Industrial & Domestic Segments ▪ The enhanced industrial market sector outputs will only marginally affect existing standards. a) The new technologies under review will have a significant

		impact upon product specifications for individual market sectors. b) New specifications will be required for the refrigerant re-use markets, the after sales markets and maintenance markets.
	▪ What NPD, innovation, invention does the market audit suggest we create for each sector and what benefits will they produce?	▪ New control system will allow separate options that are different for each segment sector – lower cost solution/more market friendly. a) Component life monitoring will create less risk for the user processes – consumer down times. b) The new refrigerant pump will close a gap in the market c) New integrated components will simplify User spare part purchasing & IVS sales processes (less component parts) d) Deletion of Product A & B for the domestic market will double the manufacturing capacity for revised product C. e) New product D will make smaller existing solar heating installations and increase efficiency by 15%. f) New product E using micro wave technology will reduce domestic heating bills by 20%

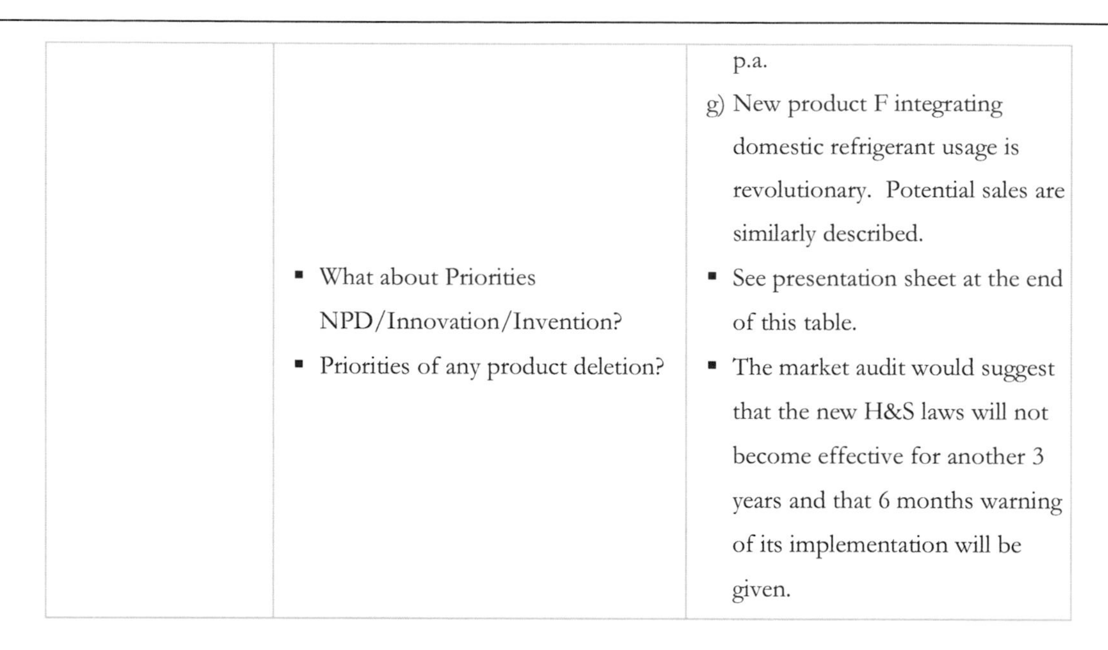

	▪ What about Priorities NPD/Innovation/Invention? ▪ Priorities of any product deletion?	p.a. g) New product F integrating domestic refrigerant usage is revolutionary. Potential sales are similarly described. ▪ See presentation sheet at the end of this table. ▪ The market audit would suggest that the new H&S laws will not become effective for another 3 years and that 6 months warning of its implementation will be given.

Table 7.6

It is suggested that product policy analysis be undertaken for each individual market sector. This will provide the basis for fine tuning the attractiveness for each sector and to then assess them in terms of segment overlap capabilities.
The main advantage of using this assessment philosophy is the eventual selling techniques that will be employed by IVS. Experience would indicate that having to sell products without any directional product policy is one of the most frustrating factors a sales person must bear (ISMM survey 2007).
It could also be argued that the same lack of directional policy prompts a need to review the discount structure more. Similarly, it could be argued that this same lack of direction is one of the factors that impact so much on the number of sales training aids that inhabit our planet.
The product policy should allow IVS to adopt a basic design philosophy that incorporates all segment needs/aspirations. Options would then be available that redefine product usage in relation to these same market sector needs. Put simply, we are maximising our opportunities by maximising market attractiveness.
The presentation sheets put forward suggested are based upon operational and long term strategic product policy.

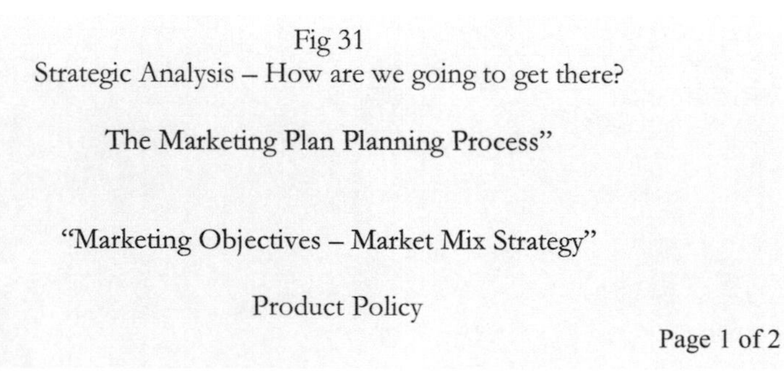

Fig 31.1
Strategic Analysis – How are we going to get there?

The Marketing Plan Planning Process"

"Marketing Objectives – Product Policy"

Operational

- Implement new control system technology
- New refrigerant pump development
- Introduce a more proactive after sales service function
- Introduce a retrofit service facility

Long Term

- Delete Product A & B for domestic market segment
- Re develop Product C for the domestic market segment
- Develop Product D for solar heating systems
- Develop Product E to reduce energy costs
- Develop integrated refrigerant philosophy
- Develop noise reduction facilities

Page 2 of 2

The directional policy undertaken is seen to create product NPD as well as a specific level of innovation/invention. The analysis provided states that we prioritise this work however we must also review resource needs and the cost of this resource. All of which will have a direct impact upon the company's profit and loss account. This would suggest that any such data pertaining to this activity be left until the following chapter (Forecasts & Budgeting). However, for the purpose of *business plan* documentation needs it is suggested that the information be produced in table format at this time (Table 7.7).

Product	Priority	No. Dev. Man Hours	Responsibility
Implement new control system technology	1	60	Engineering supported by sales & manufacturing.
New refrigerant pump development	6	300	Engineering supported by sales & manufacturing.
Introduce a more proactive after sales service function (Spare	2	250	Sales supported by accounting.

parts & maintenance)			
Introduce a retrofit service facility	3	250	Engineering supported by sales & manufacturing.
Delete Product A & B for domestic market segment	5	200	Sales supported by engineering, manufacturing and accounting.
Develop Product D for solar heating systems	4	600	Engineering supported by sales & manufacturing.
Develop Product E to reduce energy costs	7	500	Engineering supported by sales & manufacturing.
Noise reduction design	8	200	Engineering supported by sales & manufacturing.
Develop integrated refrigerant philosophy	9	900	Engineering supported by sales & manufacturing.
Total		3260 man hours	

Table 7.7

Pricing policy

How can we increase industry attractiveness with a corresponding positive impact upon the financial well being of IVS? Table 7.8 refers.

Pricing Policy	Intended Effect	Strategic Impact
Price positioning	Industrial Segment	Industrial Segment
	▪ Tendered price evaluation techniques – how do we maximise our opportunities?	▪ Compliant price is provided as a standard. IVS additional benefits are identified and priced separately together with the benefits they provide for end user and market intermediary.
	▪ Installed price evaluation techniques - how do we maximise our opportunities?	▪ Price is based on known variables, boundary definition criteria is stated to prevent confusion during the contract period.
	▪ What must we do to create price	▪ IVS quality and reliability has been

	premium for quality & reliability?	shown to create the least operating cost over the life of end user process activity.
	▪ What must we do to limit the impact upon our competitive nature?	▪ Revert back to a marketing based organisation. The net impact will be to reassert our competitive aims.
	▪ What must we do to limit the impact upon our packing and delivery costs?	▪ The market audit would suggest the potential to recycle packing material thus reducing costs.
	▪ What must we do to limit the impact upon our installation services?	▪ The market audit would suggest more research of micro environmental conditions at the time of tender thus impacting upon IVS differentiating factors.
	Domestic Segment	Domestic Segment
	▪ What is the competitive status of our products?	▪ The only increase in market share over the next 3 years will be achieved with the NPD that will impact upon the Hotel & Office building sectors.
	▪ What must we do to create price premium for quality & reliability?	▪ The market audit would suggest that only a specific percentage of the market will pay premium prices. IVS must expand its market reach.
	▪ What must we do to limit the impact upon our competitive nature?	▪ IVS business strength is low in terms of volume manufacture for low end market needs. It is suggested we continue on an operational basis pending resolution of the integrated domestic product.
	▪ How will discounts improve IVS	▪ Discounts are linked to volumes

	financial operability?	associated with manufacture and cost of material base. Increase sales will reduce manufacturing costs.
Price levels and business impact	Industrial Segment ▪ Will the new control system for existing Products A, B & C improve IVS margins? ▪ In what way will we maximise the margins for the new solar pump? ▪ In what way will the new micro wave technology impact upon the business? ▪ How will we maximise our margins on the refrigerant re use product? ▪ How will we maximise our margins on after sales? ▪ How will we maximise our margins on maintenance services? Domestic Market ▪ What impact upon margins will the deletions of Products A&B produce? ▪ What margins are expected for a revised Product C? ▪ What is the anticipated margin for the domestic refrigerant market? ▪ In what way will the overseas manufacture of the low end domestic Product C impact upon	Industrial Segment ▪ See presentation sheet at the end of this table. ▪ See presentation sheet at the en of this table. ▪ See presentation sheet at the end of this table. ▪ See presentation sheet at the end of this table. ▪ See presentation sheet at the end of this table. ▪ See presentation sheet at the end of this table. Domestic Market ▪ Industry attractiveness will reduce the margins for Product A&B over the next two years by 15% p.a. ▪ The mini survey would suggest that Product C margin will be twice the current margin in percentage terms than Product A. ▪ This is an innovative solution, only projected costs are known at this time. ▪ Significant reductions would apply to existing labour rates. But labour accounts for only 15% of total

	the P&L?	costs. Reductions in costs of unique IVS component base using overseas manufacture will be offset with the increase in testing costs to maintain quality & reliability.
Discount structure	Industrial segment ▪ Can we offer discounts for stage payments? ▪ Can we offer discounts for long term contracts? ▪ Can we offer discounts to secure a contract on a low price basis during off peak market conditions? Domestic Segment ▪ Can we offer discounts for standard products?	Industrial Segment ▪ Standard tender documents will be amended to reflect discounts applicable to stage payment. ▪ Yes by using the marketing plan to identify the benefits associated with long term contracts for growth markets in relation to macro environmental factors. ▪ The company only pursues quality enquiries. Lower prices tend to be offset when there are difficulties in comparing competitive offers on a like for like basis. Discounts are built into tendered prices for market sectors know to function in this way. Domestic Segment ▪ Discounts are offered when a product variable is no longer manufactured. The price basis is structured in relation to number of products ordered.
Market intermediaries	Industrial Segment ▪ To what extent does our pricing policy create differentiating factors for IVS? Domestic Segment	Industrial Segment ▪ By using IVS brand image the marketing intermediaries are able to differentiate their product offering. Domestic Segment

	▪ To what extent does our pricing policy create differentiating factors for IVS?	▪ The use of IVS compliments the overall mix of products available from the intermediary thus it maximises their market reach.

Table 7.8

This analysis is all about the price impact upon market attractiveness. They are the factors observed as given above that create a need for change for the existing strategies employed that will improve market attractiveness; fig 32 refers.

Fig 32
Strategic Analysis – How are we going to get there?

The Marketing Plan Planning Process"

"Marketing Objectives – Market Mix Strategy"

Pricing Policy

Page 1 of 2

Fig 32.1
Strategic Analysis – How are we going to get there?

The Marketing Plan Planning Process"

"Marketing Objectives – Pricing Policy"

Operational

- Retain existing tendering procedures but add stage payment discount structure
- Implement changes arising out of packing material recycling
- Introduce micro environmental surveys for each market sector
- Implement new pricing structures for operational NPD

Long Term

- Identify exit price strategy for Domestic Product A&B
- Define pricing structure for new Product C
- Implement new pricing structures for operational NPD
- Develop discount structure for long term contracts
- Develop integrated refrigerant philosophy

Page 2 of 2

One of the major advantages of this presentation is the commercial changes the company must undertake that IVS can use as a differentiating factor. Not what the sales process deems necessary for IVS to function. It should also be observed that IVS may already operate with differentiating factors as stated (in-built discounts) hence the need to state what commercial function must be maintained.

To some extent there exists the potential to revisit the SWOT analysis to identify or reassert the impact pricing policy has upon inter departmental needs. To this end, the factors for market attractiveness could act as drivers for a new accounting software package.

Place Policy

How can the place policy increase market attractiveness in a way that balances with the customer processes rather than imposing IVS process needs upon the market base? Table 7.9 refers.

Place Policy	Intended Effect	Strategic Impact
Product availability	Industrial & Domestic Segment ▪ Is our existing resource base able to accommodate market growth over the next 3 years? ▪ Is our existing resource base able to offset competitive activity over the next 3 years? ▪ Is our existing resource base able to accommodate the implementation of NPD & innovation over the next 3 years?	Industrial & Domestic Segment ▪ Management structure must be changed. a) Manufacturing facilities will remain unchanged. ▪ Additional staff is required to support additional marketing activity. See presentation sheet at the end of this table. ▪ Additional staff is required to undertake the R&D activity and changes in the manufacturing process associated with new products. See presentation sheet at the end of this table.
Product delivery	Industrial & Domestic Segment ▪ Do we satisfy all of the purchasing needs arising out of the market audit?	Industrial & Domestic Segment ▪ IVS currently satisfy sale channel needs with the exception of internet based after sales purchasing process requirements.
Supply logistics	Industrial & Domestic Segment ▪ How quickly do we respond to market	Industrial & Domestic Segment ▪ The audit suggests that IVS is the

	enquiries compared to competitors?	only company that confirms receipt of client documentation, have the best reception facilities and the telephone manner is much simpler than competitors.
	▪ Do we undertake a contract acceptance function?	▪ All contracts are assessed in terms of that tendered and accepted in the context of that offered and accounted for.
	▪ How do our delivery periods compare with our competitors?	▪ The audit suggests that our delivery periods do not create any differentiating factors. However the methods in which the products are despatched make them easier to install compared to our competitors.
	▪ Do we keep our customers informed of the delivery status?	▪ The audit would suggest a preference to install software that allows on line access to our process activity. This will in fact reduce the correspondence needs of IVS.
	▪ How do our warranty periods compare with our competitors?	▪ IVS standard warranty is 12 months more than competitors. Our quality and reliability allows this differentiating factor.
	▪ Do we undertake post contract communication?	▪ The market audit would suggest that IVS now undertakes this function. IVS will benefit as it will allow measurement of the benefit offered.

Table 7.9

The lack of any specific information within this policy should not be confused with the aims associated with the factors that have been identified (see fig 33). It must be remembered that these are the strategic directional factors that impact upon the *business plan*. These same factors will have more relevance within the following chapter 7.0 where we will analyse how the strategy simplifies the production of the company's profit and loss account.

Fig 33
Strategic Analysis – How are we going to get there?

The Marketing Plan Planning Process"

"Marketing Objectives – Market Mix Strategy"

Place Policy

Page 1 of 2

Fig 33.1
Strategic Analysis – How are we going to get there?

The Marketing Plan Planning Process"

"Marketing Objectives – Place Policy"

Operational

- Implement changes to management structure
- Increase competence base to offset competitive activity
- Implement R&D facilities for long term strategic needs
- Implement new on line process activity software

Long Term

- Implement changes to management structure
- Increase competence base to offset competitive activity

Page 2 of 2

Promotion Policy

How can the promotion policy increase market attractiveness in a way that maximises opportunities within the target markets? Table 7.10 refers.

Promotion Policy	Intended Effect	Strategic Impact
General comment	Industrial & Domestic Segment ▪ How does our web site analysis compare with audit results?	Industrial & Domestic Segment ▪ Our web strategy aligns closely with market attractiveness and business capabilities in that there are two major web seekers:-

		a) Those of the industrial segment seeking technical support. b) Those of the domestic sector seeking price and product details.
	▪ How does the web site data improve our market attractiveness?	▪ Market feedback indicates that our technical information as well as the documents that can be down loaded assists with end user and market intermediary planning.
	▪ Do we receive comments in our feedback box?	▪ Yes but they were never previously reviewed in terms of how they impact upon market attractiveness.
	▪ What impact upon Fig 10 above is possible by offering on line shopping?	▪ Research would indicate that this method of trading will drive forward a different range of product standards with a consequent growth of a commodity based product mix. It will not be considered as part of our strategic intent.
Advertising	Industrial & Domestic Segment ▪ Is our existing advertising effective, if so how is success measured? ▪ Do we advertise across the macro environment? ▪ What impact does our marketing have in improving our market attractiveness?	Industrial & Domestic Segment ▪ No formal marketing is undertaken at this time with the exception of several trade journal advertisements. ▪ The market audit would indicate IVS brand image has migrated across the macro environment through word of mouth. ▪ The industrial segment audit would prefer more exposure at trade shows. Both the industrial and domestic market prefers more mail shots prior to the issue of product

	▪ Should we implement a formal advertising policy?	benefit changes. ▪ Yes to reinforce the brand image prior to the launch of a change in product mix arising out of NPD & innovation.
Personal Selling	Industrial & Market Segment ▪ Which individual market sectors require a personal reintroduction to IVS and its marketing mix? ▪ Who will undertake the visit? ▪ What strategy will they employ? ▪ How will the strategy impact upon the macro environment?	Industrial & Market Segment ▪ All Market sectors; they will be prioritised in terms of growth. ▪ Manfred or Eugene plus a team member in conjunction with a senior end user and market intermediary manager. ▪ Where are they now? a) Where do they want to be? b) Can we help them get there? c) What can we do when they get there? ▪ The strategy will help to streamline NPD & innovation/invention in relation to its impact upon the macro environment.
Sales promotion	Industrial & Marketing Segment ▪ Are there any major events occurring in the UK where we could contribute to macro environmental improvements? ▪ Are there any major events that occur annually or bi-annually where we could contribute to macro environmental improvements?	Industrial & Marketing Segment The questions given hereof did not form part of the market audit and the SWOT analysis undertaken. It was agreed that this particular market potential will form the basis of a separate review following completion of this *business plan.*

Public relations	Industrial Segment ▪ Do we perceive a need to improve or otherwise our public relations? Domestic Segment ▪ Do we perceive a need to improve or otherwise our public relations?	Industrial Segment ▪ Feedback would indicate that our competitors undertake more public relations in the form of business lunches and external activities. Domestic Segment ▪ Feedback would indicate that our competitors undertake more public relations in the form of business lunches and external activities. ▪ Feedback would indicate we have become too technically orientated for the domestic market segment.

Table 7.10

Here again the lack of any specific information within this policy should not be confused with the aims associated with the factors that have been identified. Operational strategic intent will increase market attractiveness for current business activity and as such will offset competitor activity. We should however have a concern with how we will use the promotions policy to reassert our status as market leaders for future years. It is suggested that the personal selling techniques stated will create a significant differentiation factor for IVS by way of personal contact and its impact upon IVS strategic intent; fig 34 refers.

Fig 34
Strategic Analysis – How are we going to get there?

The Marketing Plan Planning Process"

"Marketing Objectives – Market Mix Strategy"

Promotions Policy

Page 1 of 2

Fig 34.1
Strategic Analysis – How are we going to get there?

The Marketing Plan Planning Process"

"Marketing Objectives – Promotions Policy"

Operational

- Reassess down loadable web data in terms of improving market attractiveness
- Implement web comments as part of the marketing process activity
- Identify a more suitable advertising policy

Long Term

- Identify suitable trade shows and their cost
- Revise the selling process to better support IVS strategic intent
- To undertake additional market research to assess target market potential (major events – Olympics?)
- Undertake a separate marketing plan for Integrated Refrigerant usage*

Page 2 of 2

*Denotes that the market audit, SWOT analysis and directional policy philosophy employed does not allow us to gauge adequately marketing technique usage. That is, we have structured our directional policy upon industry attractiveness (Fig 27) based upon existing technology usage and potential NPD.

The integrated refrigerant product would suggest a level of innovation and as such a revised audit should be issued seeking data on that already asked but also how the innovation technology can be packaged to make it attractive to potential market segments.

As suggested within the chapter notes one could expand upon the objective analysis stated within the example. For instance it could be argued that the directional policy employed refers in the main to the factors associated with a pro active strategic intent. They do not assess risk factors such as the health and safety issues that prompt the deletion of Product A&B from the example shown.

To offset any criticism that may arise out of the philosophy employed this author would suggest that any "risks" are more concerned with the implementation of the strategies and these be assessed in relation to their impact upon forecasting and budgeting as given within the next chapter.

Thus we can complete this section by creating a sheet (fig 35) to start our next chapter.

Fig 35
Strategic Analysis – How are we going to get there?

The Marketing Plan Planning Process"

"Business Plan Forecasting – Budgeting"

Page 1 of 1

The remaining portion of this chapter is used to provide supporting text for marketing strategy within a technological environment.

Supporting business plan research criteria

It will be seen that the example provided concentrates on what it is that makes a particular market attractive and how IVS reacts to this marketing consideration. Clearly, there may be no marketing undertaken whatsoever within the market macro environment and as a consequence there is potential for IVS to significantly improve their market presence.
To establish if marketing does exist within the segment there is a need to review the various marketing mixes in respect of the sub markets that may exist (Dixon and Wilkinson (1984, P.62)). These same sub markets can be considered as a "microenvironment" (Kotler et al (2002, P.118), Piercy (2003, P.69)).
Technology can give rise to several micro environmental factors for consideration, Table 7.11 refers.

Author	Factors for consideration
Beard and Easingwood (1992, P.5)	"The suggestion that the design of a marketing strategy can be based on careful research of a marketplace can seem unrealistic in high-tech markets"
Gardner et al (1998, P.1053)	"The influence of high technology is pervasive" "It affects the products and services that we purchase"
Meldrum (1994, P.45)	Does the technology used by the firm create unique marketing demands?
Moncrief and Cravens (1999, P.330)	The advance in technology enables sales individuals to be no longer territory driven, they can call on clients anywhere in the world.

Table 7.11

Methods of accommodating these factors, it is suggested cannot be achieved using strategic market planning (Beard and Easingwood (1992, P.5)). They argue a need to be more proactive than reactive

using technology push techniques. The emergence of a better-educated society during the 1990s would tend to support this argument (Zineldin (2000, P.9)).
The implementation of technology-push techniques would suggest the potential for creating unstable uses of technology by customers who are unsure of their own demands (Beckenstein and Bloch (1994, P.15)). There is a need to monitor the performance of the technology by analysing customer needs and requirements and to review test market results (Wang and Montaguti (2002, P.82)).

In the example provided the potential to expand technology, product and market needs through to new product development, R&D/marketing interface and profit as part of the overall *business plan* was postulated. Is there then the possibility to expand this postulation further? To do this we must make the assumption that the internal issues or the microenvironment represents many if not all aspects of a company's corporate plan. Support for this assumption is derived from various sources as shown in Table 7.12.

Author/s	Supporting literature
Beard and Easingwood (1992, P.5)	Consider the reasons for the potential lack of strategic marketing planning within technology intensive companies. These are given as a high level of uncertainty associated with the market.
Coates and Robinson (1995, P.14).	Show that the employment of marketing systems within a none marketing orientated company resulted in a conclusion that the application of marketing has implications for company management
Chelsom (1998)	Suggests that the technological impact upon our society creates an argument for a change in management techniques.
Hart (1999, P.14)	Concludes that strategic marketing issues where management are strategically concerned has a direct impact upon their research into product deletion strategies. These same deletion strategies impact upon the overall strategic plans of the company.
Hooley et al (2004, P.23).	Uses the same concept as McDonald to describe the principals associated with marketing fundamentals but with the added "consolidation" factor
McDonald (2003, P.42)	Argues that a corporate plan can be audited in terms of where it is now, where it wants to go and how is it going to organise its resources in order to get there.
Piercy (2003, P.394)	Stresses the importance that changes in the environment

	facing the company has upon the company's planning.

Table 7.12

As can be seen journal research authors would seem to neglect the need to structure the companies technology in respect of market attractiveness. This would suggest that these same authors presume that all of the vendors that are operating within a macro environment all function using marketing techniques. Whereas research would prove that for certain segments none of the suppliers within a given technology based macro environment undertake marketing principles (Erfurt 2002).

Forecasts & Budgeting

The success of any *business plan* in terms of business activity must be capable of being measured in financial terms. It is the definition of what is to be measured that must be observed. As we have already seen, many variables exist within the scope of the marketing plan. These same variables need not apply to "estimating expected results".
Text book data is available that could be used to provide guidance for most if not all business types within this section. It allows us to gauge the financial effect of our planning both in terms of revenue and profit. Here it can be seen that the latter are dependant upon two variables these being;

- Objective criterion
- Gap Analysis

The marketing objectives discussed in the previous chapters do not take into consideration financial details to some extent as they are based upon the strategies that need to be employed to effectively produce a *business plan.* Rather, it will be seen that it is the "Gap analysis" in an extended format that serves to create a focus for financial evaluation. In text book statement format this is given as:-

> "Once you have established the desired total revenue and profit by year over the next five years, calculate your expectations as they are now, and estimate the gap. This is called the gap analysis."
>
> Source: Davidson, H (1997, P.421) *Even More Offensive Marketing.*

We can also add to this statement by way of reverting back to the "maximising opportunities" factors within the previous sections. These are the factors borne out of the market audit, SWOT analysis and portfolio analysis (from which the gap analysis emanates) that identify opportunities that may be apparent but not currently considered. Put simply, any "gap analysis" would support the need for alternative P&L statements that this chapter will review.
The same reasoning can then be used to produce a paradigm for use as a definition for the P&L statement employed; this is given in Fig 36.

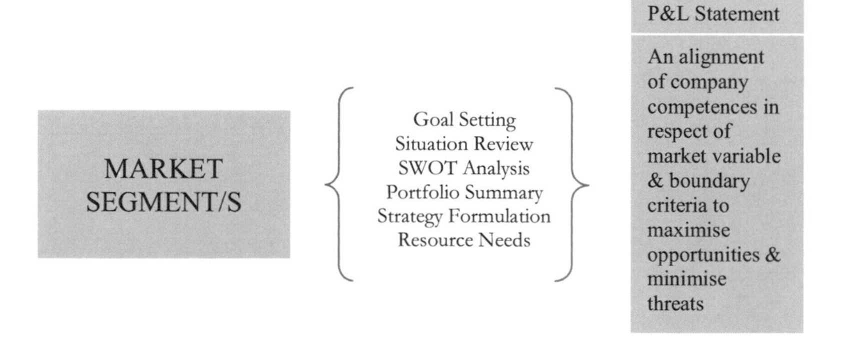

Fig 36 P&L Statement Construction Elements – Structured using a Marketing Plan

Any gap in the marketing mix portfolio will most certainly give rise to the need for some form of additional investment by way of additional funds, machinery and or manpower. It is expected that the individual responsible for producing the *business plan* will bear in mind the need to create a balance of available competences in respect of :-

- Productivity Vs Value/Volume
- Existing product/Existing market Vs Market penetration
- Existing product/Existing market Vs New products/New markets
- Value of new product Vs Development for new markets

Adapted from: McDonald, M (2003, PP.594-595) *Marketing Plans: How to Prepare Them, How To Use Them (fifth Edition).*

Here it is suggested that new or existing product development through innovation or invention will impact more upon the expected financial results. These in turn will have a congruent impact upon any market segment under review. Therefore, it is argued that those employing marketing techniques will be better positioned to undertake any product development with limited risk association as they will be in a better position to identify a gap analysis for their company.

It is to be noted that the management of any gap analysis has the potential to attract danger in terms of the marketing objectives undertaken and the strategies employed to implement them. Resource allocation is perhaps the activity that creates the highest risk. Clearly, any eagerness to undertake a specific new development or innovation can be quickly tempered if the resources are not available locally or nationally to implement them. Moreover, if using marketing intermediaries who do not possess marketing skills, they may not have the facilities in their own business activities that sufficiently allow them to position or differentiate your business within their business activity. Let us also not forget the competitive nature of the services required as an external resource need.

In summary there may be a gap between what business owners and managers want and the cost of what the *business plan* is suggesting. There exists therefore the potential for contingencies to be borne in mind right from the outset of the marketing planning process.

Here it is suggested that business owners and managers refer back to the corporate objectives chapter and the investment assessment criteria put forward. This can be summarised in the following way:-

- In what way will the objectives employed impact upon the total projected sales, the cost of sales and the margins they generate.
- **In what way will any "gap analysis" employed impact upon the total projected sales, the** cost of sales and the margin generated.

The author would advocate that any other methods used to produce a P&L statement will require more focus on the management of the selling techniques undertaken in order to achieve the expected yearly results. Cut price sales or additional discounts may be required just to meet anticipated targets in particular cash flow projections as discussed further within this chapter.

This being said, there is also a need to structure a profit and loss account where the marketing plan and its output is shown to impact upon the business. In particular, the accuracy of the projected volume of sales and the times of the year they occur.

We can begin to examine the products in terms of the directional policy matrix in the context of their target markets. We can measure any impact upon the business caused by changes in the marketing mix in the context of the investment undertaken. All of which can be measured in respect of operational as well a long term business objectives. But in what way will this all impact upon the methodology employed to construct accurate forecasts and budgets?

8.1 Profit and Loss Statement Construction Techniques

Much work has already been undertaken in the previous sections that will allow one to readily project a profit and loss statement, hereinafter called the P&L. This being said, why leave the P&L until the last when presenting ones *business plan*. If left until the last the presenter risks criticism from others if the projected financial summary is not what is expected or hoped for.
Furthermore, the remaining sections of the *business plan* presentation will contain data on how the P&L has been constructed and thus there exists the potential for others to suggest ways of improving its content. To this end, there may be ways of enhancing the financial performance of the business.

Irrespective of where the P&L is located within the presentation, personal preference of those reviewing and/or approving the plan will need to be assessed prior to the completion of the plan.

Every *business plan* must consist of a P&L statement. It is normal practise to provide an operational P&L, namely that P&L which is used to reflect the status of the business for the current or the next fiscal or tax period. Similarly, there will be many senior managers or financial institutions who may dictate a review of the long term strategy for the business under review. This being so, the individual producing the *business plan* will need to assess if the long term strategy reflects a three or five year period. The corporate objectives would normally determine the useful periods that are used for assessment.

It is to be noted that ongoing businesses will have previous trading activity. Ergo, one should also list the financial results of the previous year. Growth or otherwise can then be projected as a comparison with previous (actual) trading activity.

The structure of the P&L statement will vary from company to company. If a large company, the accounts department together with the company secretary will produce a document that meets the needs of the tax and auditing fraternity as well as those of the shareholders if operating as a public company. Small to medium size businesses may be reliant upon external accounting institutions to produce the P&L for them. Here, the term "accounting" reflects adequately what it is they are undertaking. That is, what money is coming into the business, how much of the money is used to produce the marketing activity and how much is retained and how will it be used.

Any money coming into the business will be as a result of a "Sale". Here it must be remembered that the word "Sale" is not what most people associate with the term "Order". Sales persons take orders whereas it is the business that delivers the "Sale". Indeed, a sale is only complete upon receipt of payment for the product or service in question.
It follows that order book values for a company will be different to the projected sales of a company. Accordingly, we can state that the P&L would consist of "Total Sales" and the "Cost of Sales". The difference between these two items would be the "Profit" generated by the sale.

We can expand upon these three definitions but this will become more evident as we continue. Thus we now have a reason for a P&L statement together with a basis of understanding of how it should be constructed.

8.1.1 Total Sales Analysis

Firstly, total sales for any company may consist of a mix of different products or services that have or will be traded over a certain period. It is at this point of the *business plan* (right at the outset) where many companies fail to project correctly their true strengths or competences. Nor, it could be said do these same companies substantiate the use of these competences in respect of market influences and their impact upon the operational and long term strategic financial well being of the company. Put simply, the P&L may not be all that it seems to those approving its application.

The following financial analysis of company's total sales as shown in table 8.1.1 will provide the basis for examining such a comment in more detail.

1	2	3	4	5	6	7	8
Product/Service Type	Target Market	Price Each	Quantity	Total Sale	Cost Each	Total Cost	Segment margin
A	A	£1000.00	200	£200,000	£50.00	£10,000	£190,000
B	C	£1500.00	120	£180,000	£100.00	£12,000	£168,000
C	B	£2000.00	100	£200,000	£130.00	£13,000	£187,000
D	B	£2500.00	180	£450,000	£200.00	£36,000	£414,000
			Totals	**£1030K**		**£71K**	**£959K**

Table 8.1.1

It can be seen that the total sales of the company is derived using a number of different products or services that are sold to different market segments. Here we need to assess two important factors.

- The product or service mix in terms of operational projections
- The product or service mix in terms of long term strategic projections.

The work we have undertaken this far should allow us to construct any number of tables depending upon the corporate objectives we are endeavouring to provide a *business plan* for. Perhaps more importantly, the factors we now have at hand by way of marketing plan application should provide more realism to the P&L statements put forward as shown in Fig 37.

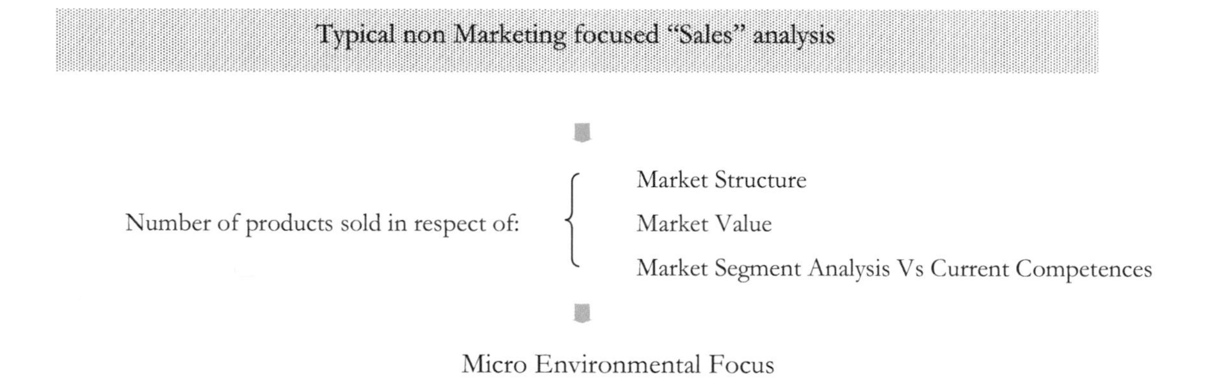

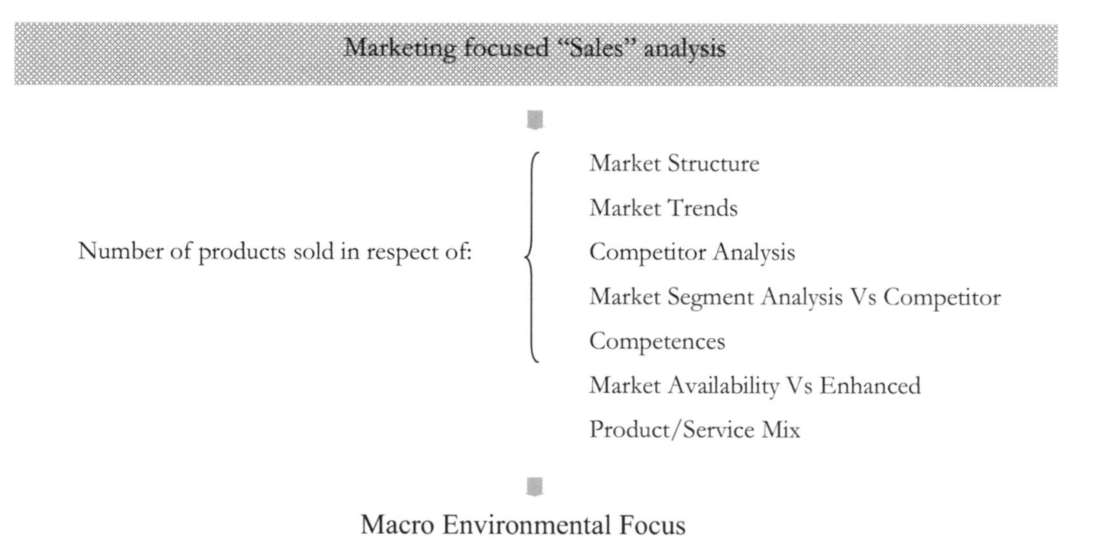

Fig 37 Sales Analyses Factors

This comparison shows that non marketing focused organisations may concentrate more on providing statistical market data that is more associated with those who purchase the company's product or service to produce the P&L. Typically, it is tantamount to having a product or service and to then endeavour to quantify potential buyers and their impact upon total sales. Yes, this type of plan may suffice for the budding new inventor or entrepreneur keen to maximise the sale of their market offering.
However, as stated within previous chapters this same offering will have a life cycle and as such what must be done for the business to survive pending its reduction in market demand and/or the advent of competitors?

> Here we must not forget our main aim of maximising our opportunities and minimising our risks. Clearly, any focus associated with monitoring only those who purchase or who will purchase the market offering does not maximise sales opportunities

This is the objective of the sales department not the business as a whole. That is, "Sales" as a department has responsibility of achieving order intake whereas it is "Marketing" as a department or consultancy service that must create and balance the *business plan* between corporate objectives, company departmental needs in respect of market opportunities.

8.1.2 Cost of Sales Analysis

The cost of sales and the manner in which its value is presented has evolved in several ways. Larger organisations will need to comply with mandatory methods of presentation as dictated by auditing authorities and their subsequent compliancy requirements of others. Whereas small to medium business institutions may provide, bespoke to them, alternative methods of cost of sales value representation.

Irrespective of the method in which a company's P&L is produced it should be remembered that our criteria for the *business plan* must be the projected marketing strategy and the cost of its implementation. If the accounts department have a specific preference in the manner in which this statement is tabulated then so be it. We must however not detract from the methodology associated with the manner in which the cost of sales is identified.

Cost of sales is a factor within any business where every employee has a direct impact upon the financial well being of any company. This statement can be put into another format whereby cost of sales can be subdivided into direct and indirect costs.
A better understanding of the latter makes it easier to construct a *business plan.*

8.1.2A Direct Costs

Direct costs are those costs associated with the production of a given product or service. For instance, if the business has its own manufacturing plant the direct costs would be the cost of material

and labour associated with producing the product. Ergo, the costs depicted within column 7 Table 8A would be these same costs.
If the business employs a production manager then he or she must be closely involved in any analysis of direct costs. If operating without such a manager the business owner as well as the person producing the ***business plan*** must be able to understand and assess the strategic importance that direct costs have on the business.
Factors for consideration are numerous but for ease of depiction the main such factors (those that readily impact upon the selling price) are tabulated as given in Table 8.1.2. Prudence would dictate that any design department as well as the accounts department review the same factors as will become obvious.

Table 8.1.2 Direct Costs – Analysis & Comments	
Cost Description	Comment

Material Costs	*Own Manufacture*
	• Identify which materials are crucial to the market offering. That is, the materials that are classed as a commodity and those that may be bespoke.
	• In what way do these materials come together to produce the market offering (see chapter summary)?. List the commodity and bespoke items separately as a percentage of the total material cost (see chapter summary).
	• Why are these materials required? List in terms of product features associated with (see chapter summary):
	  - Basic product design
	  - Marketing requirement (derived from marketing plan)
	  - Quality requirement
	  - Sales Channel requirement (discussed further in)
	  - Other specific criteria
	• At what stage of the life cycle is the material component at in respect of the market offering (see chapter summary)? Note that this is an important aspect of this part of the plan.
	• Identify source of origin for each product type (commodity & bespoke).
	• In what way are they purchased noting their impact upon company cash flow and capital employed? Here it may be prudent to list separately the commodity items, particularly those that are bought as

Material Costs Continued

stock. If bought as stock the accounts department will need to assess how their costs are amortised across the product base.

- How are bespoke and/or those items with a high monetary value purchased? What process is used to limit the negative impact upon cash flow during its "Work in Process" life cycle? Can you allow finance charges as well as discounts for order down payments within the overall cost base (see indirect overheads chapter).
- In what way are the material costs "accounted" for each product type to allow actual product costs to be monitored.
- What is the reject rate for the manufacturing plant? State in terms of monetary value and also as a comparison with previous years. State the reason for the reject rate and what work is ongoing to reduce any values quoted.
- How is the cost of material rejects accounted for? Is it as a direct cost or indirect?
- What materials are needed to support the manufacturing activity as a whole (not necessarily for one product type). Does the volume used warrant producing one's own production of this need.
- What is the overall cost of the latter and how has such cost's evolved over a given period?

Bought In Materials

Here one should assess all of the above bullet points as a method of evaluating any suppliers used to supplement the cost of sales. Important factors for consideration are:-

- For the major cost items and perhaps the bespoke items state the process used to identify the most suitable product supplier.
- Name the major suppliers and state why they have been selected.
- State the names of alternate suppliers and their impact upon cost of sales in the event that the major supplier is unable to source the goods ordered.
- If using large quantities of raw materials (steel, copper, oil, gas, timber etc) list how the cost of the commodity has evolved and

<table>
<tr><td></td><td>if possible cost and supply projections for the future.

In the event that either of the above factors have been a cause for concern for previous trading years it is advisable to have separate presentation material at hand that identifies what will be done to improve the situation.</td></tr>
<tr><td>Labour Costs</td><td>Own Manufacture
Here it is important to assess accurately the number of man-hours required to produce a product type. That is, what is the manufacturing capacity of the company? The importance originates from the simple formula put forward.

No. of Products Available = Number of Manufacturing Hours Available X Manufacturing Efficiency

For example if one person is employed for 40 hours per week but only 30 hours of the week are actually applied to the production process the number of hours available is only 30 hours. Therefore, if a product is estimated at requiring 120 hours to produce it then it will take four weeks to complete the process. This would then imply that 13 products would be produced within a 52 week period.

Use this same formula (or similarly derived) for the operational P&L as well as any long term planning assessment.

• Which factors impact the most upon efficiency? – list separately. It could be:-

– The type of tools and/or machinery is used within the production process i.e. how do they impact upon the man hours expended?
– What is the cost of preparing the machine for multiple manufacture of component parts? Can this cost be reduced and if so what would be its impact upon the cost of sales.</td></tr>
</table>

	- The ergonomics of the production environment. - List any other factors that the company has to pay for.

Table 8.1.2

It is at this stage of the business plan that the subject of globalisation may give rise to the use of an alternate cost base. Many companies have now transferred their manufacture to areas where the cost of labour is significantly lower than within their own geographic regions. Yet, companies such as Ferrari still make cars in Italy!
So in what way do we assess this potential paradox?

We must first complete our marketing plan and *business plan* activities and review our conclusions before changing our area of focus. We may have already considered this aspect within the strategies we wish to employ as discussed within chapter 7.0.

8.1.2B Indirect Costs

These types of business costs are indirect because they are not directly attributable to the sales process, but to what extent and with what degree of unpredictability? They are costs that may not be needed to produce a product and as such they are accounted for separately. In this way, they can be more easily amortised across the overall cost of sales.

There is no rule of thumb that can be used to define exactly the type of indirect costs applied to a specific business type or industry type. The factors that determine them are dependant upon the industry that one operates within and the business strategy employed by the organisation.

They can be listed or grouped into specific areas depending upon how the accounts department wishes to identify them. Similarly, it will become evident that some of the factors listed will in fact attract a direct cost but the cost of administration must be accounted for. The company accounts officer or outside accountant will suggest the best way of cost identification. Typical indirect costs are shown in Table 8.1.3.

Table 8.1.3 Indirect Costs – Analysis & Comments	
Cost of Staff	Salaries for all staff. - Provide total headcount for the business and list the

<table>
<tr><td></td><td>departments they are allocated to.<ul><li>List total cost for each department types separately e.g. Production, Sales & Accounting.</li><li>Cost of fringes e.g. NH & Pension and Medical Contributions for each department type.</li></ul>The foregoing three bullet points should not be calculated on what the business can afford. There must be a direct relationship in respect of man-hours over the yearly period to provide the necessary marketing effort required to produce the volume of sales.
That is, what are the total sales envisaged in respect of manufacturing and labour capacity. Clearly, a level of trade-off will be required to balance the overall cost of staff.</td></tr>
<tr><td>Cost of Utilities</td><td>Electricity, Water, Gas, Telephones.<ul><li>Irrespective of usage identify mix of users and costs. Provide an analysis of departmental usage.</li><li>Let each department supervisor account for the level of expenditure undertaken. Indeed, what steps are they taking to effectively manage the most efficient method of controlling costs – chapter 7.0 strategic actions?</li><li>Large users of a specific type of energy source or communications provider should review any impact upon cost of overheads created by investment in alternate methods of self generation.</li></ul></td></tr>
<tr><td>Engineering and Design (E&D)</td><td><ul><li>If an order requires some form of E&D to produce the product the cost of that E&D should be identified in reference to the E&D needs of the projected total sales for the year/s.</li><li>There is a need to note the E&D man - hours available during the year and its amortised availability for the number of orders projected.</li><li>If allocating the E&D cost on a per job basis then the appropriate cost (possibly in a man hour rate) should be identified separately to the cost of staff given above. That is, if the cost of E&D staff is calculated on a per contract basis then the</li></ul></td></tr>
</table>

	E&D staff costs should not form part of the *cost of staff* calculation. That is the E&D is financed by prospective orders. Additional notes are provided within the example forming part of this chapter
Research and Development	This particular indirect cost has the potential to attract much debate as well as substantial expenditure. The main criteria that must be observed with any R&D expenditure is that the work undertaken and its cost must be shown to have a direct impact upon sustaining and growing total sales, for instance:- • New products for new markets. • New products for existing markets. • Creating new markets for existing products. • Enhancing the existing product mix for existing markets. • Cost reduction exercises • Quality Improvements • Reliability improvements • Increased or more efficient methods of manufacture The R&D expenditure undertaken for any or all of the above bullet points should be presented in the form of the R&D impact upon the P&L statement as determined by the marketing plan. It is possible that several P&L documents may need to be produced to show the net result of each or a grouped activity. The structure of the P&L should be based upon the overall impact upon total sales plus the indirect cost changes needed to support the appropriate expenditure and implementation.
Packing costs	What is the cost of packing the product? Here prudence would dictate that the total for the year is estimated so that it can be amortised across each product. If a specific department is employed to undertake this requirement it may be prudent to amortise only the material costs and to include the labour costs required to manage the function within the cost of staff.
Delivery costs	It is possible that market conditions require that the selling price includes delivery. This being the case there may be a direct cost involved as well

	as the cost of administration. If a specific department is employed to undertake this requirement it may be prudent to amortise only the actual costs and to include the labour costs required to manage the function within the cost of staff.
Maintenance	All aspects of the business will require some form of maintenance. For larger companies specific planned maintenance may be undertaken at a given time of the year in order to sustain production activity. Smaller companies may require lights to be changed or repairs to the boiler or air conditioning system. If planned maintenance is involved then a specific sum must be included within this section. Plus the provisional sum that would normally be set aside by the small to medium business to keep the business active. Attention to Health & Safety procedures as well as any other occupational heath requirements must be observed. Any costs associated with this requirement should be allocated accordingly.
Company Cars	Certain levels of management as well as sales staff will attract as part of their remuneration package a company car of a specific type. There may also be added benefits in the form of fuel allowances. The method of cost identification will need to be discussed with the finance department as the cost of the asset can be amortised over a number of years depending upon the government taxation rules prevailing at any given time.
Fixed assets	Equipment will be required to produce a means to an end be it for a specific direct or indirect cost factor. It could be manufacturing machinery or office computer based equipment respectively. Similarly, an office or building will be required from which the business will operate. Here again the method of cost identification will need to be discussed with the finance department as the cost of the asset can be amortised over a number of years depending upon the government rules prevailing at any given time.
Professional Services	Most companies will require an outside company to undertake some form of professional service. Clearly, small to medium sized businesses will require more services than the larger companies. Accounting and legal services are two such requirements that spring to mind.

	It is suggested that cost of professional services be broken down into two sections. One section where the professional services to sustain business activity is listed as a separate cost. • One section where professional services are required to expand the business (e.g. marketing consultancy) is listed as a separate cost. Here again the second bullet point would be used to augment any alternate P&L statements produced as part of the *business plan*. Prudence would suggest that some form of contract be agreed between the business and service provider where the cost of service is identified as a yearly amount thus simplifying the costing and accounting procedures.
Insurance costs	Insurance costs as a percentage of sales for a small to medium organisation can be much higher than that for a large company. This particular item has two connotations in that the primary requirement is to ensure the cost of all insurances is allowed for within the P&L. The secondary function is that the *business plan* will allow prospective insurers to better identify more aspects of the business thus ensuring a more accurate insurance premium.
Stationary Costs	All companies will need to correspond internally and externally. The amount of stationary required to fulfil such a need will vary from company to company. If the costs are considered high it is suggested that the total yearly cost be broken down into individual departmental usage. There onwards any necessary action can be taken to define more efficient methods of stationary usage.
Advertising	The cost of advertising is dependant upon the level and type of advertising perceived to support the business. Cost of brochures, mail shots and any internet activities as well as promotional gifts all need to be

	accounted for. Here it is important to state why the advertising and methods employed to undertake it are required. Actual reasons for advertising can be gleaned from the remaining sections of the business plan. Perhaps more importantly, however is the impact that advertising will produce upon the P&L statement.
Training	Training can be assessed in many ways. Irrespectively, all costs borne by the company must be retrieved and accounted for accordingly.
Any other indirect cost	The indirect costs shown above are typical. Any other indirect cost (sales process improvements?) specific to a given industry type could be used to expand upon those given.

Table 8.1.3

It is important to note several key bullet points during the process of calculating indirect costs for the business.

- At all times the costs identified must be seen to support the business activity. If additional staff or machinery is required then their implementation must coincide with a direct impact upon total sales. For example the purchase of a new manufacturing machine will reduce labour costs by a given amount and/or will produce more of a specific product during the year.

- Where indirect labour is required to fulfil a given task prudence would dictate that the relevant departmental supervisor assess the man hours associated with the task from which the need for labour emanated. If the hours assessed are estimates then these should be listed within the *business plan* stating the basis for which the estimate is undertaken and what measures will be used to gain a more accurate assessment.

- There will always be a potential to package separately some of the indirect costs for any business. For instance, certain orders or contracts may be specific in such a way that the cost of a given indirect activity may form part of the overall selling price of a given product or service offering.

Following completion of the assessment of direct and indirect costs we can produce an intermediary P&L analysis. Using the data derived above this can be depicted in a simple format as shown in Fig 38.

Note that table 8.1 column 7 has now been amended to show "Total Direct Cost".

1	2	3	4	5	6	7	8
Product/Service Type	Target Market	Price Each	Quantity	Total Sale	Cost Each	Total Direct Cost	Segment margin
A	A	£1000.00	200	£200,000	£50.00	£10,000	£190,000
B	C	£1500.00	120	£180,000	£100.00	£12,000	£168,000
C	B	£2000.00	100	£200,000	£130.00	£13,000	£187,000
D	B	£2500.00	180	£450,000	£200.00	£36,000	£414,000
			Totals	**£1030K**		**£71K**	**£959K**

Less Indirect costs

Cost of Staff	£200K
Cost of Utilities	£50K
Engineering and Design (E&D)	£100K
Research and Development	£100K
Packing costs	£15K
Delivery costs	£20K
Maintenance	£30K
Company Cars	£75K
Fixed assets	£75K
Professional Services	£30K
Insurance costs	£5K
Stationary Costs	£5K
Advertising	£10K
Training	£10K
Net Margin	**£234K**

Fig 38 Intermediary Profit & Loss Statement

The term intermediary has been used to define the amount of profit generated within the projected Total Sales Vs Total Cost analysis depicted in Figure 38. This is because we have yet to review the cost of financing the business.

All businesses will require monetary funding of some description to sustain its business activity. Substantial wording is available that supports the importance of finance/cash availability for the business. For the purpose of this book we will review the impact that money creates when assessing the *business plan* and its presentation to others. Bullet points are used to simplify the explanation of the appropriate requirements.

- If cash is not available such that the business requires an overdraft or loan then there will be bank charges to account for. Accordingly, the finance department should advise the appropriate amount to be included as an indirect cost.

- Similarly, if cash is available but is used to finance the operation of the business then interest will be lost because the cash could be used to generate interest.

- Many organisations operate by way of invoicing for the sale. They then have to await payment from the buyer before depositing the money in the bank. Unless payment is prompt, say within the month of invoice submittal then bank charges or loss of potential interest would apply and provisions may be (should be) provided.

- Many organisations operate by way of payment via a debit/credit card system. Here there will be a charge for the point of sale equipment as well as a handling fee.

- Product or Service price in relation to market activity may be such that allowances need to be made in the form of provisions for any bad debts risks.

Suffice it to say that all business may or will encounter difficulties arising out of any or all of these bullet points. For the purpose of the *business plan* presentation it is suggested that all fixed costs, namely bank charges and point of sale charges be estimated for the yearly period and added to the P&L as an indirect cost.
At the same time, if the company operates within an environment where payment is known to extend beyond 30 days then statements as to methods of control should form part of the finance department individual performance business plan submittal. Here it is suggested that a separate finance charge be calculated for each tender or sale as an indirect cost and included within the overall selling price.
The total sum of all the finance charges that impact upon the business throughout the year should be shown as a separate line at the bottom of the intermediary P&L as depicted within figure 38.

The last two items for consideration to be shown at the base of the P&L are projected taxation monetary amounts as well as any share holder dividends. Clearly any net profit derived from the business will need to be assessed after taxation. Similarly, any profit retained will be determined following shareholder dividend issue.
Both of these items have the potential to be complex both in the manner they are calculated as well as the manner in which they are reflected as part of the *P&L*. For the purpose of simplification, it is suggested that methodology associated with projecting taxation and share holder dividend monetary values be left to the last. Even then, it is suggested that the onus of completing this particular section be left entirely to the finance department and the company secretary.

8.2 Budgeting

All businesses are structured such that they have a capacity to respond to the market environment. Certain environments operate on a constant basis where the daily or monthly output of the business operating for 365 days throughout the year is adequate. Other environments are seasonal such that only a few months of the year are considered as months of operational activity.
Irrespective of the activity associated with any given business there is a need to undertake a formal budget analysis for inclusion within the *business plan*. The contents of which can become complex without a formal review process.
It is suggested that the budget analysis phase of the plan consists of the following strategic needs.

- Cost of implementing the *business plan*
- Operating income

The cost of implementing the *business plan* is to a large extent determined by the gap analysis and the strategies employed to modify the marketing mix. The P&L statement put forward will show that the costs attributable to any gaps can be accrued and reconciled over the yearly period through our sales efforts. Indeed, one need only assess the myriads of software programmes that are available that are used to monitor sales performance. Sadly, however these same projections rarely if ever compare these projections to operating income.

Businesses cannot be maintained if they require loans or overdrafts on a continuous basis. Yes they may be required to support the business during lean periods. It is the money derived through our sales processes that must be used to sustain business activity. Moreover, we need to project accurately at what periods during the year sales derived income will be achieved. That is, we must project our cash flow.

Product Type	June			July			August		
	Qty	Sales	Margin	Qty	Sales	Margin	Qty	Sales	Margin
Product A Target Market A	40	1600	1400	50	2000	1750	50	2000	1750
Product A Target Market C	20	840	740	30	1260	1110	30	1260	1110
Total		2440	2140		3260	2860		3260	2860

Table 8.2

Table 8.2 is used as an example of cash flow projections for a three month period. Here we can project the amount of monies received for a given month and compare it with our overall business costs for a given month. It follows that our finance department can gauge the accrual of available margin as operating income and its impact upon the company bank statement. It can also be used as the basis to forecast a loan or overdraft.

Having at hand the information provided by Table 8.2 will also provide additional valuable support for the business in several ways.

- The company may employ several sales individuals who all sell a mix of products or services. Without table 8.2 how are they able to assess which potential business opportunities have priority over others? Similarly, how will they know how important the accuracy of budgeting impacts upon the business?

- If a certain product produces 70% of the turn over (Cash Cow or Star) one will be able to gauge at any one time how it supports the business. Similarly, the "Dog" can be identified in relation to other products as well as its impact upon the business.

- You may be operating with a limited capacity for supply to the extent that the need's of a low margin market is detracting from the ability to produce high margin product? Do you need to revisit the SWOT analysis as well as marketing mix to create a differently balanced mix portfolio?

- How are market trends/performance factors impacting upon cash flow? Do you need to revisit the strategy employed, that is buck or boost the strategy. Here it should also be noted that marketers are interested in markets. They will often asses what is termed the "Pareto" (80/20 rule) effect within which 20% of the market produces 80% of the business and that this same rule applies to the remaining 80% of the market and so forth. It is strongly recommended that even without the marketing plan that that business owners assess repeat business and that they check if this applies to their business, particularly cash flow projections.

8.3 Chapter Summary

Clearly, the forecasts and budgets chapter provided serves to create a better financial foundation and understanding of any given business where no such understanding or limited understanding previously existed.
It allows us to use our marketing plan to project a more accurate financial assessment of the cost of the strategy employed to meet corporate objectives. Congruently, it provides us with the triggers to determine strategy continuation or strategy re-analysis. Clearly, financial statistics will need to be undertaken by the business. Statistics it must be said should be those that will allow the business to better respond to market influences.

For example, financial performance can impact upon market performance and vice versa. Donaldson, H (1997) provides us with interesting data to support this comment, namely:-

- "A 5% increase in annual customer retention can increase total company operating profits by over 50%.

- Most companies lose 25% of their customers annually yet most companies spend six times as much on generating new customers as on retaining existing ones"

The need to provide charts or tables as part of the ***business plan*** was alluded to previously within this chapter. In table 8.1.2 reference was made in which the response format should, as a minimum be considered as follows.

Material costs:	Here there is a need identify as a percentage of the total cost of sales the material cost per product per target market possibly in pie chart format. This will allow managers to assess the potential impact upon margin arising out of changes in raw material costs. Provide a separate chart showing the individual costs of materials such as steel, copper etc as well as outsourced products. This will allow managers to assess the impact on total cost each material has.
Material Life Cycle:	How long has specific material types/outsourced products been used? At what stage of their life cycle are they considered to be at? Is further development required to extend the life cycle? Probably most suited to a table type of presentation
Cost of Utilities:	Here one need only analyse the percentage of overall usage by each department possibly in pie chart format. For instance, telephone based communication seems low for the sales department considered to the production department.

Chapter example

Given that the overall objective of this chapter is to produce a suitable profit and loss statement we can begin this example by identifying where we are at in terms of the marketing plan in relation to figure 36 given above.
It will be noted that we have yet to identify resource needs in respect of the strategic intent (inclusive of our gap analysis) of the business as described within the previous chapter. Furthermore there is a need to balance these same resource needs with maximising our opportunities and minimising our risks for operational and long term objective needs.

Before we can begin our resource analysis there is a need to examine the data associated the four bullet points linked with figure 11 given above. The draw back with any analysis at this juncture is that the first bullet point "*Productivity Vs Value/Volume*" does not have any variable definition nor is there a stated boundary definition associated with them.

However, given that our task is to analyse and provide a suitable and acceptable profit and loss statement we could argue that the P&L is the boundary definition criteria we must employ and that variable analysis is dependent upon those factors that form the basis of the P&L. It follows that boundary definition will vary for each company as will the extent to which one wishes to construct the P&L.
For this example we will start by analysing projected sales for each product type for each market segment in relation to manufacturing resource and its association with product cost as depicted within our chapter P&L analysis. The market has been assessed in respect of operational activity as well as that projected for IVS 3 year long term objectives. The analysis is based upon the completion of the new control system development to offset competitive activity.

Operational Projections – Industrial Market Segment

Steel Industry

Product	Market Value	IVS Share	Projected Sales (Qty)	Price (each)
A	2 million	60%	12	100K
B	1.5 million	62%	15	62K
C	1 million	65%	12	54K
Retrofit	300K	100%	1 (yearly contract)	300K
After Sales	200K	N/A	2 (provide list of parts)	200K
Solar Pump	N/A	N/A	N/A	N/A
Maintenance	70K	N/A	1 (yearly contract)	70K
M/Wave	N/A	N/A	N/A	N/A
Refrigerant Integration	New audit required	New audit required	New audit required	New audit required

Table 8.3

Food & Beverage

Product	Market Value	IVS Share	Projected Sales (Qty)	Price (each)
A	2 million	61%	12	102K
B	1 million	48%	7	64K
C	1 million	61%	11	56K
Retrofit	400K	100%	1 (yearly contract)	400K
After Sales	150K	N/A	2 (provide list of parts)	150K
Solar Pump	N/A	N/A	N/A	N/A
Maintenance	75K	N/A	1 (yearly contract)	75K

M/Wave	N/A	N/A	N/A	N/A
Refrigerant Integration	New audit required	New audit required	New audit required	New audit required

Table 8.4

Paper & Pulp

Product	Market Value	IVS Share	Projected Sales (Qty)	Price (each)
A	3 million	32%	10	95K
B	1 million	22%	4	55K
C	2 million	24%	10	48K
Retrofit	1 million	0%	N/A	N/A
After Sales	75K	N/A	2 (provide list of parts)	75K
Solar Pump	N/A	N/A	N/A	N/A
Maintenance	30K	N/A	1 (yearly contract)	30K
M/Wave	N/A	N/A	N/A	N/A
Refrigerant Integration	New audit required	New audit required	New audit required	New audit required

Table 8.5

Petro - Chemical

Product	Market Value	IVS Share	Projected Sales (Qty)	Price (each)
A	5 million	53%	22	120K
B	3 million	58%	25	70K
C	2 million	51%	17	60K
Retrofit	600K	N/A	1 (yearly contract)	600K
After Sales	70K	N/A	2 (provide list of parts)	70K
Solar Pump	N/A	N/A	N/A	N/A
Maintenance	60K	N/A	1 (yearly contract)	60K
M/Wave	N/A	N/A	N/A	N/A
Refrigerant Integration	New audit required	New audit required	New audit required	New audit required

Table 8.6

Operational Projections – Domestic/Light Industry Market Segment

Urban & Sub Urban

Product	Market Value	IVS Share	Projected Sales (Qty)	Price (each)
A	1.5 million	20%	300	1K
B	2.5 million	25%	1000	625
C	1,5 million	10%	300	500
Retrofit	N/A	N/A	N/A	N/A
After Sales	90K	N/A	100 (provide list of parts)	90K
Solar Pump	N/A	N/A	N/A	N/A
Maintenance	N/A	N/A	N/A	N/A
M/Wave	N/A	N/A	N/A	N/A
Refrigerant Integration	New audit required	New audit required	New audit required	New audit required

Table 8.7

Note that Product A, B & C for this market segment are similar to the industrial segment but exclude all of the external hardware associated with industrial segment needs.

Light Industry (Hotels, Office Buildings & Major Stores etc)

Product	Market Value	IVS Share	Projected Sales (Qty)	Price (each)
A	3 million	30%	25	40K
B	2 million	21%	12	35K
C	2 million	20%	13	30K
Retrofit	N/A	N/A	N/A	N/A
After Sales	90K	N/A	50 (provide list of parts)	90K
Solar Pump	N/A	N/A	N/A	N/A
Maintenance	N/A	N/A	N/A	N/A
M/Wave	N/A	N/A	N/A	N/A
Refrigerant Integration	New audit required	New audit required	New audit required	New audit required

Table 8.8

Note that Product A, B & C for this market segment are similar to the industrial segment but incorporate substantially less external hardware than the industrial segment needs.
It is suggested that readers resist the temptation to expand significantly upon these tables at this time. Here it must be remembered that IVS has to align its resource accordingly in terms of manufacturing

resource and human competency (Overheads). Put simply we need to align "*Productivity Vs Value/Volume*"

Long term Projections – Industrial Market Segment

Projections based upon the completion of all actions arising out of the market mix directional policy, marketing priorities/critical success factors and segment growth have been sourced internally. It could be argued that the tables given be amended to reflect sales growth associated with the potential export market. However, such data is excluded in order to retain a level of simplicity for this example.

Steel Industry

Product	Market Value	IVS Share	Projected Sales (Qty)	Price (each)
A	3 million	66%	18	110K
B	2 million	65%	20	65K
C	1.5 million	72%	18	60K
Retrofit	100K	100%	1 (yearly contract)	100K
After Sales	100K	N/A	2 (provide list of parts)	100K
Solar Pump	100K	100%	10	1K
Maintenance	75K	N/A	1 (yearly contract)	75K
M/Wave	N/A	N/A	N/A	N/A
Refrigerant Integration	New audit required	New audit required	New audit required	New audit required

Table 8.9

Food & Beverage

Product	Market Value	IVS Share	Projected Sales (Qty)	Price (each)
A	2.5 million	66%	15	110K
B	1.5 million	56%	12	70K
C	1.5 million	64%	16	60K
Retrofit	100K	100%	1 (yearly contract)	100K
After Sales	150K	N/A	2 (provide list of parts)	150K
Solar Pump	70K	100%	7	1K
Maintenance	80K	N/A	1 (yearly contract)	80K
M/Wave	N/A	N/A	N/A	N/A
Refrigerant Integration	New audit required	New audit required	New audit required	New audit required

Table 8.10

Paper & Pulp

Product	Market Value	IVS Share	Projected Sales (Qty)	Price (each)
A	2 million	35%	7	100K
B	2 million	24%	8	60K
C	1 million	44%	8	55K
Retrofit	N/A	N/A	N/A	N/A
After Sales	75K	N/A	2 (provide list of parts)	75K
Solar Pump	80K	100%	8	1K
Maintenance	30K	N/A	1 (yearly contract)	30K
M/Wave	N/A	N/A	N/A	N/A
Refrigerant Integration	New audit required	New audit required	New audit required	New audit required

Table 8.11

Petro - Chemical

Product	Market Value	IVS Share	Projected Sales (Qty)	Price (each)
A	7 million	55%	30	130K
B	4 million	60%	30	80K
C	3 million	58%	25	70K
Retrofit	200K	N/A	1 (yearly contract)	200K
After Sales	90K	N/A	2 (provide list of parts)	90K
Solar Pump	150K	100%	15	1.5K
Maintenance	70K	N/A	1 (yearly contract)	70K
M/Wave	N/A	N/A	N/A	N/A
Refrigerant Integration	New audit required	New audit required	New audit required	New audit required

Table 8.12

Operational Projections – Domestic/Light Industry Market Segment

Urban & Sub Urban

Product	Market Value	IVS Share	Projected Sales (Qty)	Price (each)
C	7 million	15%	1500	700
Retrofit	N/A	N/A	N/A	N/A
After Sales	120K	N/A	120 (provide list of	120K

			parts)	
Solar Pump	1.5 million	60%	600	1.5K
Maintenance	N/A	N/A	N/A	N/A
M/Wave	800K	100%	1000	800
Refrigerant Integration	New audit required	New audit required	New audit required	New audit required

Table 8.13

Note that Product A, B & C for this market segment are similar to the industrial segment but exclude all of the external hardware associated with industrial segment needs.

Light Industry (Hotels, Office Buildings & Major Stores etc)

Product	Market Value	IVS Share	Projected Sales (Qty)	Price (each)
C	8 million	41%	100	33K
Retrofit	N/A	N/A	N/A	N/A
After Sales	150K	N/A	250 (provide list of parts)	150K
Solar Pump	1 million	36%	120	3K
Maintenance	N/A	N/A	N/A	N/A
M/Wave	400K	100%	100	4K
Refrigerant Integration	New audit required	New audit required	New audit required	New audit required

Table 8.14

These tables reflect the opportunities that exist for IVS arising out of the marketing plan undertaken thus far. To this end, these are the sales that IVS could readily expect to achieve in accordance with our strategy formulation.

We now need to assess our resource base in a manner that will balance the use of our resource both for operational needs and for long term objective requirements.
At this time our interest should be limited to manufacturing resource. The overhead resource base analysis will follow later.

Manufacturing resource for this example will take into account available labour and machinery resource and the production environment. This is further explained using the following simplistic analysis:-

Operational Requirements

Manufacturing resource – Industrial & Light Industry Segment
(Values based upon one manufacturing shift)

	Yearly production capacity		
	Product A	Product B	Product C
Machine X	120	-	-
Machine Y	-	90	-
Machine Z	-	-	90
Test Plant A	110	100	100
Packing M/C A	140	120	120
Fork Lift A	140	130	130
Projected sales*	81	67	63

Table 8.15 *Denotes additional row has been added to compare with projected sales.

Manufacturing resource – Domestic Segment
(Values based upon one manufacturing shift)

	Yearly production capacity		
	Product A	Product B	Product C
Machine X1	400	-	-
Machine Y1	-	3000	-
Machine Z1	-	-	2000
Test Plant A	500	3500	2500
Packing M/C A	600	3500	2500
Fork Lift A	450	3500	3500
Projected sales*	300	1000	625

Table 8.16 *Denotes additional row has been added to compare with projected sales.

Long term objective requirements

Manufacturing resource – Industrial & Light Industry Segment
(Values based upon one manufacturing shift)

	Yearly production capacity		
	Product A	Product B	Product C
Machine X	120	-	-
Machine Y	-	90	-
Machine Z	-	-	90
Test Plant A	110	100	100
Packing M/C A	140	120	120
Fork Lift A	140	130	130
Projected sales*	70	70	167

Table 8.17 *Denotes additional row has been added to compare with projected sales.

Manufacturing resource – Domestic Segment
(Values based upon one manufacturing shift)

	Yearly production capacity				
	Product A	Product B	Product C	Solar Pump	M/Wave
Machine X2	-	-	-	-	2000
Machine Y2	-	-	-	1000	-
Machine Z1	-	-	2000	-	-
Test Plant A	500	3500	2500	-	-
Test Plant B	-	-	-	1500	1800
Packing M/C A	600	3500	2500	2000	2000
Fork Lift A & B	450	3500	3500	2000	2000
Projected sales*	Nil	Nil	1500	720	1100

Table 8.18 *Denotes additional row has been added to compare with projected sales

Prudence would suggest that we tabulate how our manufacturing resource meets with projected operational and long term objective sales potential. Similarly, the tables provided will be seen to impact upon the *business plan* the manufacturing resource will produce as will be seen in Table 8.19 below.

Machine/equipment	Operational Needs/Benefits	Long Term Needs/Benefits
Existing plant	▪ Existing equipment utilisation factor is not	▪ No additional shift patterns are envisaged.

	greater than 75% ▪ Machine Z1 utilisation factor is 30% (Product C)	▪ Machine Z1 utilisation factor will increase to 75%
Existing plant to be deleted	▪ Machines X1 & Y1 will need to be shut down for a period of 4 weeks.	▪ Low utilisation factor will allow a build of stock during machine X2 & Y2 installation period.
Existing Test Equipment	▪ The utilisation factors can be maintained for 18 months.	▪ New diagnostic equipment at a cost of 10K will need to be purchased.
New Test Equipment	▪ Existing test equipment can be upgraded at a cost of £20K to accommodate new products.	▪ New test equipment will require an additional $10M^2$ of factory floor space.
Fork Lift Equipment	▪ New H&S laws mandate a change from gas operated to electrical operation at a cost of 5k. ▪ Existing fork lift utilisation will accommodate operational plus long term except the micro wave product.	▪ Factory gas extraction fans no longer required. ▪ New fork lift at 15K will require an additional battery charger which needs a new building extension.
Stores & Warehouse Facility	▪ Existing warehouse undergoes overfill during peak periods.	▪ Long term projections indicate an additional $50M^2$ manufacturing space.

Table 8.19

The information provided within this table would suggest that separate research be undertaken into expanding the manufacturing facilities. For ease of explanation we will assume that the facility is rented from the local municipality who have granted IVS a lease on the adjoining property and that the latter will be used as the warehouse.

It must be borne in mind that the analyses given above in respect of operational needs compared to long term object requirements refer only to the manufacturing capacity envisaged. Indeed, we could suggest that we are entering a phase of the business plan that would prompt the analogy "*The Plate Spinning Part of the Business Plan*".

To this end, what is the cost of the machinery, what is it impact upon the direct and indirect overheads and perhaps more importantly, in what way does it impact upon the margin associated with the projected sales prices? Will IVS have to consider a change in selling price if so, what impact will this have on projected sales quantities?

Yes there are many variables we must still assess but for the purpose of arriving at an integrated planning process several of the tables given above have been shaded as presentation material based upon the facts known at this time. They will prove useful later within our total manufacturing cost analysis (direct & indirect costs) as will the following manufacturing labour resource analysis.

Operational Requirements
Manufacturing labour resource – Industrial & Light Industry Segment

	Labour Hours per product		
	Product A	Product B	Product C
Machine X	6	-	-
Machine Y	-	5	-
Machine Z	-	-	4
Test Plant A	2	2	2
Packing M/C A	1	1	1
Storekeeper (amortised)	1	1	1
Assembly	3	3	2
Wiring	5	3	4
Fork Lift A (amortised)	1	1	1
Warehouse personnel (amortised)	1	1	1
Total	20	17	16
Projected sales	81	67	63
Total man hours p.a.	1620	1139	1008

Table 8.20

Manufacturing resource – Domestic Segment
(Values based upon one manufacturing shift)

	Yearly production capacity		
	Product A	Product B	Product C
Machine X1	70mins	-	-
Machine Y1	-	50mins	-
Machine Z1	-	-	45mins
Test Plant A	10mins	10mins	10mins
Packing M/C A	2mins	2mins	2mins
Storekeeper (amortised)	20mins	20mins	20mins
Assembly			
Wiring	15mins	12mins	10mins
Fork Lift A (amortised)	5mins	5mins	5mins
Warehouse personnel (amortised)	10mins	10mins	10mins
Total	132mins	107mins	102mins
Projected sales	300	1000	625
Total man hours p.a.	792hrs	1784hrs	1063hrs

Table 8.21

Given that we will need to assess our resource needs as they are now in relation to that projected by way of increased sales and NPD it is suggested that we identify at this time our manning level resource needs.

Operational

Segment	Product A	Product B	Product C	Total
Industrial	1620	1139	1008	3767
Domestic/	792	1784	1063	3639
Total	2412	2923	2071	7406

Table 8.22

Resource requirements are:-

Man hours per person (potential) p.a. = 9000

Manufacturing efficiency = 80%

Man hours per person (available) p.a. = 7200 (average)

Total Manufacturing Hours = 7406 (operational)

No of staff needed = 7406 / 7200

= 1.03

The practicability of this assessment is questionable as clearly, one person cannot operate the whole of the manufacturing resource base for this example. The resource is therefore more dependent upon staffing levels that are needed to "set up" and operate the machines. It follows that we must be careful not to provide data for analysis that would restate that already put forward (tautology). Our long term analysis for this example (using our manufacturing resource data analysis) shows that we are removing two machines and adding two more machines with no real increase in production output. It is for this reason that a long term objective analysis is not provided. All that is required is a statement alluding to the fact that no increase in manufacturing labour resource is needed to fulfil long term objectives.

It will be noted that we are not following any given chronological process to arrive at our aims. This will always be dependent upon the individual responsible for preparing the *business plan.* One good reason for this is the reliance upon others to procure the data needed to complete the plan. What we can say however is that our primary objective is to identify and tabulate all of the factors required to fulfil the needs of table 8A. This is the table that lists the direct costs required to produce a product.

A review of the process undertaken this far will show that labour costs and material costs must be identified before we can produce our direct cost analysis (Fig 8A)

We have already identified the number of staff required for the manufacturing facility to function. This is given as Table 8.23:-

Operational		Long Term	
Machine X & X1	2	Machine X & X2	2
Machine Y & Y1	2	Machine Y & Y2	2
Machine Z & Z1	2	Machine Z & Z1	2
Test Plant A	1	Test Plant A	1
Packing M/C A	1	Packing M/C A	1
Storekeeper	1	Storekeeper	1

Assembly	2	Assembly	2
Wiring	2	Wiring	2
Fork Lift A	1	Fork Lift A & B	2
Warehouse personnel	2	Warehouse personnel	3
Total	16 (144Khrs)	Total	18 (162kHrs)

Table 8.23

So in what way do we account for the labour cost expended for each product? Here our accounting fraternity will have an array of potential solutions. Furthermore it is possible that specific software may be employed to undertake this task. This being so rather leave well alone unless there is a distinct need to change it. However, let us not forget the potential for software programmes to be scientific in nature such that they define costs down to a decimal point. Since when did pennies impact upon a selling price (low cost commodity items excluded)?

What must be borne in mind is the need to monitor the labour cost during the sales process activity. It follows that prudence would suggest some form of congruency in cost price analysis and the sales process accountancy methods employed. The need to refer to a company standard should act as a driving factor. That is, some form of standard the company can refer to in order to measure performance.

To all intents and purposes the labour analysis provided thus far within this example would function as a standard. Here again let us forget any specific software package that may be available and review how we an account for labour resource cost base?

For this example we will offset individual product labour usage in an amortised fashion in relation to the total manufacturing labour resource base. In formula format this is given as:-

$$\text{Product labour Allocation} = \text{Total labour} \times \frac{\text{Actual Product labour}}{\text{Product Labour}}$$

Thus we have the following allocation of the number or hours per product on a per annum basis. The cost of the labour is shown using a rate of £10/hr.

Operational

Industrial market

Product A

$$\text{Product labour Allocation} = \frac{144\text{khrs}}{7.4\text{khrs}} \times 1.14\text{khrs} = 22.18\text{khrs} = 22.18\text{K}$$

Using this same analysis for the remaining products we are able to derive the following respective labour analysis values, Table 8.24 refers.

Operational			Long term		
Industrial & Light Industrial Segment			Industrial & Light Industrial Segment		
Product A	22.18khrs	22.18K	Product A	19.17khrs	21.87K
Product B	31.53khrs	31.53K	Product B	32.95khrs	39.87K
Product C	19.66khrs	19.66K	Product C	52.1khrs	57.31K
Totals	73.37khrs	73.37K	Product C	104.22khrs	119.05K
Domestic Segment			Domestic Segment		
Product A	15.37khrs	15.37K	Product C	39.65khrs	43.62K
Product B	34.64khrs	34.64K	Microwave	9.065khrs	9.97K
Product C	20.62khrs	20.62K	Solar Pump	9.065khrs	9.97K
Totals	70.63khrs	70.63K	Totals	57.78khrs	63.56K
Totals	144.00khrs	144K		162.00khrs	182.61K

Table 8.24

The values shown within the long term objectives were assessed by revising the operational manning level resource needs by way of the ratio between new Product C in relation to old Product A&B sales. Similarly the solar pump and microwave sales are each the same by way of quantity; hence the values given.

Labour rates have been increased in accordance with projected yearly cost of living increase at 3% per annum. We will complete our labour analysis in way that will allow its relevant inclusion within table 8.1.1; Table 8.25 refers.

Operational					
Segment	Product	Total Product	Total Hrs	Unit Hrs	Unit Labour Cost
Industrial	A	81	22.18khrs	274	2,740
Industrial	B	67	31.53khrs	471	4,710
Industrial	C	63	19.66khrs	312	3,120
Domestic	A	300	15.37khrs	51	510
Domestic	B	1000	34.64khrs	34.64	346.4
Domestic	C	625	20.62khrs	33	330

Table 8.25

Long Term					
Segment	Product	Total Product	Total Hrs	Unit Hrs	Unit Labour Cost
Industrial	A	70	19.17khrs	274	3,014
Industrial	B	70	32.95khrs	471	5,181
Industrial	C	167	52.1khrs	312	3,432
Domestic	C	1500	39.65khrs	26.43	291
Domestic	Micro wave	720	9.065khrs	12.62	141
Domestic	Solar Pump	110	9.065khrs	82.41	906

Table 8.25.1

It should be noted that the unit labour rates given are based upon the man hours expended to produce a given product. They exclude overhead recovery which many of you will be familiar with. Overhead recovery rates cannot be assessed until we have produced our P&L statement in its entirety. Furthermore, let us not forget that the impact of overhead recovery serves to provide a more simple method of business expense accounting and for tendering purposes. They do not impact upon the production of statistical data at this juncture. For this reason this particular subject is discussed at the end of this chapter example.

We should now be in a position to begin our analysis of the material costs used within the production element of the sales process. Using our chapter notes given above we should concern ourselves with the cost of stock material, cost of manufactured components and cost of bought in components. Stock material is common to a range of products and as such we should amortise the total yearly cost in relation to its percentage usage across the product mix.

The cost of manufactured components and bought in components will be heavily reliant upon the specifications produced for a given product. Specifications it should be said would be those produced by the engineering or similar department that create a standard for a given product. For this example we will presume that material cost base used is sourced upon product standards that have been modified to reflect the strategic intent of the *business plan.*

Given that a range of component providers may be used by a given company who may each supply at a different price, the product standard employed may not reflect the lowest cost option for a particular product. Here it is suggested that discussions with the company accountant be held to arrive at a satisfactory solution. However, it must be borne in mind that the standards employed should reflect the cost of a product for a specific market segment. Any product features used to create a difference between market segment needs should be listed separately. That is they should not form the basis of our material cost benchmarks we will establish as part of the *business plan.* The benefits to IVS of this action are numerous particularly in measuring performance and its P&L impact caused by changes within market segment activity.

This being said the example provided does not delve into any specific product features for simplification purposes. It is given in table format as follows:-

Operational – Base Material Cost Price					
Segment	Product	Stock Material	Manufactured Material	Bought In Components	Unit Material Cost
Industrial	A	1.5k	30k	5k	36.5k
Industrial	B	1.2k	25k	4k	30.2k
Industrial	C	1.0k	20k	3k	24k
Light Industrial	A	550	12k	2k	14.55k
Light Industrial	B	500	10k	1.5k	12k
Light Industrial	C	450	8k	1.2k	9.65k
Domestic	A	20	100	40	160
Domestic	B	10	60	20	90
Domestic	C	5	20	5	30

Table 8.26

Long term – Base Material Cost Price					
Segment	Product	Stock Material	Manufacture d Material	Bought In Components	Unit Material Cost
Industrial	A	1.7k	32k	6k	39.7k
Industrial	B	1.4k	27k	5k	33.4k
Industrial	C	1.2k	21k	3.5k	25.7k
Light Industrial	A	600	13k	2.2k	15.8k
Light Industrial	B	550	11k	2k	13.55k
Light Industrial	C	500	8.5k	1.5k	10.5k
Domestic	C	22	110	44	176
Domestic	Micro wave	12	180	22	214

Domestic	Solar Pump	7	160	15	182

Table 8.26.1

We can amend the latter to include labour cost by way of tables 8.27 & 8.27.1 and thus we will be able to complete the unit cost price for inclusion within a typical table 8.1.1.

Operational – Unit Cost Price				
Segment	Product	Unit Labour Cost	Unit Material Cost	Unit Cost
Industrial	A	2,740	36.5k	39.24k
Industrial	B	4,710	30.2k	34.91k
Industrial	C	3,120	24k	27.12k
Light Industrial	A	2,740	14.55k	17.29k
Light Industrial	B	4,710	12k	16.71k
Light Industrial	C	3,120	9.65k	12.77k
Domestic	A	510	160	670
Domestic	B	346.4	90	436.4
Domestic	C	330	30	360

Table 8.27

Long term – Unit Cost Price				
Segment	Product	Unit Labour Cost	Unit Material Cost	Unit Cost
Industrial	A	3,014	39.7k	42.71k
Industrial	B	5,181	33.4k	38.58k
Industrial	C	3,432	25.7k	29.13k
Light Industrial	A	3,014	15.8k	18.81k
Light Industrial	B	5,181	13.55k	18.73k
Light Industrial	C	3,432	10.5k	13.92k
Domestic	C	291	176	466
Domestic	Micro wave	141	214	355
Domestic	Solar Pump	906	182	1088

Table 8.27.1

We now have all of the data required to complete table 8.1.1 for this example. That is, what are the total sales we will generate and what will be the manufacturing cost; Table 8.28 & 8.29 refers. The analysis provided would suggest that each market segment should have its own unit cost calculation. Here it will be seen that the light industrial market segment has the same labour and material cost as the industrial market which to all intents and purposes should vary accordingly. For this example we will suggest that the difference in selling price is caused by a difference in overhead costs such as engineering.

Operational

1	2	3	4	5	6	7	8
Product/Service Type	Target Market	Price Each	Quantity	Total Sale	Cost Each	Total Cost	Segment margin
A - Steel	1	100k	12	1200k	39.24k	470.88k	792.12k
A - F&B	1	102k	12	1224k	39.24k	470.88k	753.12k
A - P&P	1	95k	10	950k	39.24k	392.4k	557.6k
A - PC	1	120k	22	2640k	39.24k	863.28k	1777k
B - Steel	1	62k	15	930k	34.91k	523.65k	406.35k
B - F&B	1	64k	7	448k	34.91k	244.37k	203.7k
B - P&P	1	55k	4	220k	34.91k	139.64k	80.36k
B - PC	1	70k	25	1750k	34.91k	872.75k	877.25k
C - Steel	1	54k	12	648k	29.13k	349.56k	298.44k
C - F&B	1	56k	11	616k	29.13k	320.43k	295.57k
C - P&P	1	48k	10	480k	29.13k	291.3k	188.7k
C - PC	1	60k	17	1020k	29.13k	495.21k	524.8k
A	2	40k	25	1000k	18.81k	470.25k	529.75k
B	2	35k	12	420k	18.73k	224.76k	195.24k
C	2	30k	13	390k	13.92k	180.96k	209.04
A	3	1k	300	300k	670	201k	99k
B	3	625	1000	625k	436.4	436.4k	188.6k
C	3	500	300	150k	360	108k	42k
			Totals*	**15,011K**		**7,056K**	**7,955K**

Table 8.28

1 denotes industrial segment
2 denotes light industrial segment
3 denotes domestic segment

Long Term

1	2	3	4	5	6	7	8
Product/Service Type	Target Market	Price Each	Quantity	Total Sale	Cost Each	Total Cost	Segment margin
A - Steel	1	110K	18	1980k	42.71k	768.8k	1211.2k
A - F&B	1	110k	15	1650k	42.71k	640.7k	1009.3k
A - P&P	1	100k	7	700k	42.71k	299k	401k
A - PC	1	130k	30	3900k	42.71k	1281.3k	2618.7k
B - Steel	1	65k	20	1300k	38.58k	771.6k	528.4k
B - F&B	1	70k	12	840k	38.58k	463k	377k
B - P&P	1	60k	8	480k	38.58k	308.6k	171.4k
B - PC	1	80k	30	2400k	38.58k	1157.4k	1242.6k
C - Steel	1	60k	18	1080k	29.13k	524.3k	555.7k
C - F&B	1	60k	16	960k	29.13k	466.1k	493.9k
C - P&P	1	55k	8	440k	29.13k	233k	207k
C - PC	1	70k	25	1750k	29.13k	725.2k	1024.8k
C	2	33k	100	3300k	13.92k	1392k	1908k
C	3	700	1500	1050k	466	699K	351k
Micro wave	3	800	1000	800k	355	355k	445k
Solar Pump	3	1.5k	60	90k	1088	65.3k	24.7k
			Totals*	**22,720K**		**10,150K**	**12,570K**

Table 8.29 *denotes that values have been rounded.

4 denotes industrial segment
5 denotes light industrial segment
6 denotes domestic segment

If we adopt the same presentation philosophy employed throughout the latter two tables should be presented in shaded format as they will form part of the *business plan.* That is, they would not be tucked away somewhere within the appendix based supporting data.

To all intents and purposes we could conclude this particular chapter example sub section however, the sales that IVS has projected also include a retrofit and maintenance market and after sales business activity. Whereas the retrofit and after sales business activity would suggest some form of manufacturing activity the maintenance business activity will most probably relate more to personnel needs. It follows that the two former business activities must have a separate table 8.1.1 example and that these values could be added to latter above table.

To this end, as with most business activities they will utilise the same indirect overhead resource base to pursue an end to an aim.

The potential drawback associated with this activity is the possible difficulty that may arise out of performance measurement. Furthermore, there exists a probability that a separate sales process may be needed to maximise IVS market attractiveness alignment needs. It could be argued therefore that a separate P&L statement be produced; utilising the same indirect overhead resource base for these three additional business activities.

This will become more apparent within our assessment of the indirect overhead resource base. Similarly, we will refrain from presenting our sales/cost/margin factors for the three additional business activities accordingly.

We will use table 8.1.3 given within this chapter as the basis for assessing our indirect overheads for this example. It should be noted that our SWOT analysis suggested a change in management structure. Furthermore, our corporate objectives seek to create more empowerment within the organisation all of which will have an impact upon the cost of indirect overheads. The main reason for this emanates from the need to let Eugene and Manfred step back from the customer/IVS sales process interface

Cost of staff
First we will review the cost of indirect overhead needs associated with operational sales projections and the impact upon the business; table 8.30 refers;

Operational

Department	Quantity	Cost	Fringes
CEO + PA	2		
Totals		320k	64k
Sales (Manfred)			
Secretary	1		
Contracts Engineers	2		
Sales assistant	2		
New Sales Executive	1		
Total	6	420k	63k
Production (Hayley)			
Secretary	1		
Buyer	1		
Buying assistant (new)	1		

Industrial Engineers	2		
New Draughtsman	1		
Maintenance staff	2		
Total	7	550k	83k
Engineering (Jonathan)			
Engineers	2		
New Engineers	1		
Draughtsman	1		
Total	4	380k	57k
Accounting (Patrick)			
Secretary	1		
Clerks	3		
Total	4	280k	42k
Total cost of staff = 2195k			

Table 8.30

The analysis provided should follow to some extent the philosophy employed for the manufacturing resource base calculation in that indirect overhead cost should have a level of business application congruency.

For instance, the cost of "engineering" personnel has been shown as an indirect overhead. If analysed further this type of engineering for IVS is used to undertake engineering work that IVS must produce as part of the sales process. Does it not follow that we exclude such costs at this time and treat them as a direct cost?

The answer is it all depends upon the opinions of business owners and managers and the accounting processes employed to identify and measure performance. If already operating with a high percentage overhead prudence would suggest its reduction or reallocation by way of direct cost allocation. Clearly, market down trends and a consequent reduction in orders would suggest high staff turnover if applying the latter value analysis techniques.

There is no universal set of guidelines that will provide direction for any opinions in this regard. We could suggest however that there are factors that will create a more simple method of analysis for that someone who has the final say; that is, what will influence his or her opinion? Here it is recommended that we first review the product standards already alluded to.

In what way does our resource competence exert its differentiating influence within the product standard? Moreover, in what way does this resource competence become an inherent part of the product standard features irrespective of whether the differentiating feature is specified or ordered by the customer?

If quality and reliability form the basis of differentiating factors it is highly probable that engineering skills were used to produce a better competitive specification which would suggest an argument for indirect overhead allocation to prevail for this example.

Long term

Department	Quantity	Cost	Fringes
CEO + PA	2		
Totals		380k	76k
Sales (Manfred)			
Secretary	1		
Contracts Engineers	2		
Sales assistant	2		
New Sales Executive	2		
New Contracts Engineer	2		
New Sales assistant	1		
Total	10	700k	105k
Production (Hayley)			
Secretary	1		
Buyer	1		
Buying assistant (new)	1		
Industrial Engineers	2		
New Draughtsman	1		
Maintenance staff	2		
New Draughtsman	1		
Upgrade Buying assist’	1		
Total	10	700k	105k
Engineering (Jonathan)			
Engineers	2		
New Engineers	1		
Draughtsman	1		
New draughtsman	1		
Total	5	480k	72k
Accounting (Patrick)			

Secretary	1		
Clerks	3		
New Junior Accountant	1		
Total	5	320k	48k
Total cost of staff = 2986k			

Table 8.31

Cost of Utilities
Actual cost of usage was determined by the accountant (Patrick) for this example, who is the responsible person for administering payment for the cost of Utilities. Owing to turbulence within the gas and electricity pricing market the revised values shown were suggested to accommodate known market fluctuations.

Operational (yearly totals)

Department	Electricity	Gas	Telephone
Executive suite	200	-	100
Sales dept	250	-	1200
Manufacturing dept	1000	900	30
Engineering dept	100	-	300
Accounts dept	150	-	250
Totals	1700	800	1880
Revised estimates	2200	1200	1880
Total = 5280			

Table 8.32

Long Term

Department	Electricity	Gas	Telephone
Total all departments	4300	1440	2562
Total = 8302			

Table 8.33

Engineering & Design

Here we need to separate the cost of E&D other than the cost of the resource labour requirements. The latter is included within the "cost of staff" element given above. It is the cost the business will encounter through engineering design to support the sales (and marketing) process that are not encountered as part of a contract requirement.

If we separate all of the sales process elements with this departmental activity we are left with the equipment E&D required in order to operate effectively. This could be a specific range of test equipment or items of plant required to produce test pieces. It is also any special computer based equipment used to produce sales process documentation requirements. For this example we must produce a list of equipment that is seen to benefit the sales process as well a supporting the company marketing effort but not any special R&D equipment.

This could take the form of:-

Operational			
E&D Purchase Needs	Sales Process Activity	Marketing Process Activity	Cost
Larger Autocad m/c	Produces better quality diagrams at twice the rate of existing equipment	Directional policy employed will require an update of all engineering documentation	5k
Separate engineering IT server	Production process is unaffected by company IT problems.	Sensitive design data is more safely stored	10k
Purchase of new printer to replace rented unit.	Service provider cannot support production needs.	Directional policy employed will require an update of all engineering documentation	15k
Replacement of Test plant diagnostic equipment	Able to test more product C components p.a.	Existing software employed will become defunct within 14 months	30k
Sub Total = 60k			
Engineering hours	Develop new purchasing sub contract documentation	Prepare new product standards	10k
Retain existing links with local University	Increase the knowledge base	New technology awareness	20k
Sub Total = 30k			

Table 8.34

Long term			
E&D Purchase Needs	Sales Process Activity	Marketing Process Activity	Cost
Second Autocad m/c	Production demand criteria in respect of increased product output.	Additional capacity will support R&D test needs for NPD.	8k
Engineering IT server Capacity update	Production demand criteria in respect of increased product output.	Additional capacity will support R&D test needs for NPD.	12k
Sub Total = 20k			
Engineering hours	Implementation of new test plant	Additional capacity will support R&D test needs for NPD.	10k
Retain existing links with local University	Increase the knowledge base	New technology awareness	22k
Sub Total = 32k			

Table 8.35

The material cost for this overhead will need to be reviewed by the accounts department. This is because taxation rules normally do not permit the total costs of such items to be accounted for in a given years P&L statement. They are classed as having a specific useful business life for taxation purposes in terms of years. For this example we will amortise the costs as shown in Table 8.36.

Period	Year 1	Year 2	Year 3	Year 4	Year 5
Operational					
Material	12k	12k	12k	12k	12k
Labour	30k	-	-	-	-
Long term					
Material	-	-	4k	4k	4k
Labour	-	-	32k	-	-

Table 8.36

This indirect overhead has the potential to be the most frustrating to both identify and account for. The reason for this is that the level of technology employed is normally only known to the engineer seeking to purchase the new plant. By having a separate sales and marketing process column as stated

will prompt the relevant individual to be more specific in relation to the needs of the plant to warrant its purchase.
The tables given above could suggest by way of the factors stated that the technology needs of all of the other departments should also be listed at this time within this example. However, it is unlikely that the equipment associated with these remaining departments will impact upon the marketing process. Yes they may improve the sales process but at what cost impact. *Business plan* scrutinisers will always be on the look out for unwanted luxuries. Any such equipment for this example is defined within "any other indirect costs" sub chapter discussed later.

Research and Development

The directional policy employed as part of the marketing plan will result in a diverse mix of R&D needs across a mix of segments. For this example preference would indicate a need to restate what the R&D consists of in presentation format. If anything it will remind *business plan* readers what they are without having to revert back to previous sections. This is given in a simple table (8.37).

Product Policy	
Operational	Long term
▪ Implement new control system technology ▪ New refrigerant pump development ▪ Introduce a more proactive after sales service function ▪ Introduce a retrofit service facility	▪ Delete Product A & B for domestic market segment ▪ Re develop Product C for the domestic market segment ▪ Develop Product D for solar heating systems ▪ Develop Product E to reduce energy costs ▪ Develop integrated refrigerant philosophy ▪ Develop noise reduction facilities

Table 8.37

Quite a task it would seem. Perhaps more importantly in what way can we assess the cost arising out of these development objectives? We could ask Patrick (Engineer) what his assessment would be but what process would he use to arrive at an acceptable cost base? Similarly, in what way could we all be sure that provided would be within acceptable limits?

R&D attracts significant text book coverage as well as creating a basis for journal research in a direct or indirect format. One could literally spend many hours analysing a different mix of text objectives. For this example we will use the following analysis (inclusive of innovation/invention criterion) to determine our cost base (Kotler, P; et al (2002, P.501)).

But first however let us examine how research journals have the ability to impact upon the adopted process; table 8.38 refers.

Here it will be seen the difficulty technology research individuals may face and the potential solutions available to them.

Author	Relevant Journal Analysis
Cahill & Warshawsky (1993. P20)	Suggest that we let the engineers and programmers come up with the innovations but then let the potential customers have a say in whether the innovation is viable.
Czuchry & Yasin (1999. P.240)	State that a significant portion of today's technical innovation is attributed to small to medium organisations.
Thomas & Tymon (1992. P347)	Academic research principles adapted for business use. "Goal relevance refers to the correspondence of outcome (or dependent) variables in a theory to the things the practitioner wishes to influence"
Coates & Robinson (1995. P.12)	Provide supporting evidence that there are many definitions of what constitutes NPD. Their analysis was not "previously included in its catalogue" They go on to screen NPD for a company operating without a formal marketing structure. They suggest:- 1. Strategy 2. Idea generation 3. Assessment 4. Business plans 5. Development 6. Testing 7. Commercialisation
Nijssen & Lieshout (1995. P39)	They review a more expanded NPD process model but conclude that in principle an NPD model by way of its content will be of benefit to users. However, research

	would show NPD model usage in differing chronological formats. They go on to suggest that problems are seen to occur when external companies undertake market research.
Maile & Bialik (1988. P.58)	Here we have another method of NPD selection criteria. The model employed would suggest that marketing managers are able to establish NPD selectivity levels for individual product NPD and the entire product mix.
Meldrum (1994)	Compares suggested differences in technology definition criteria as well as the macro environmental association with technology. Highly suggested as recommended reading for technology based companies.
Gardner, Johnson, Lee & Wilkinson (1999. P.1057)	Review the need to compare NPD definitions for new products in a different manner to NPD for reformulated products.
Dunn, Hulak & White (1999. P.186)	Suggest a new approach to high-tech marketing segmentation. They state companies should consider:- 1. Specialised solution 2. Customised solution 3. Value Solution 4. Packaged solution (adapted for this sub chapter as an area of focus for NPD strategy & idea formulation techniques)
Grantham (1997. P.9)	Product life cycles and the sequences they consist of may not hold true for all technologies; they can experience second lives.

Palmer & Williams (2000. P.15)	Provide us with a successful application of the Fisher Pry (technology change analysis) as applied to computer microprocessors.

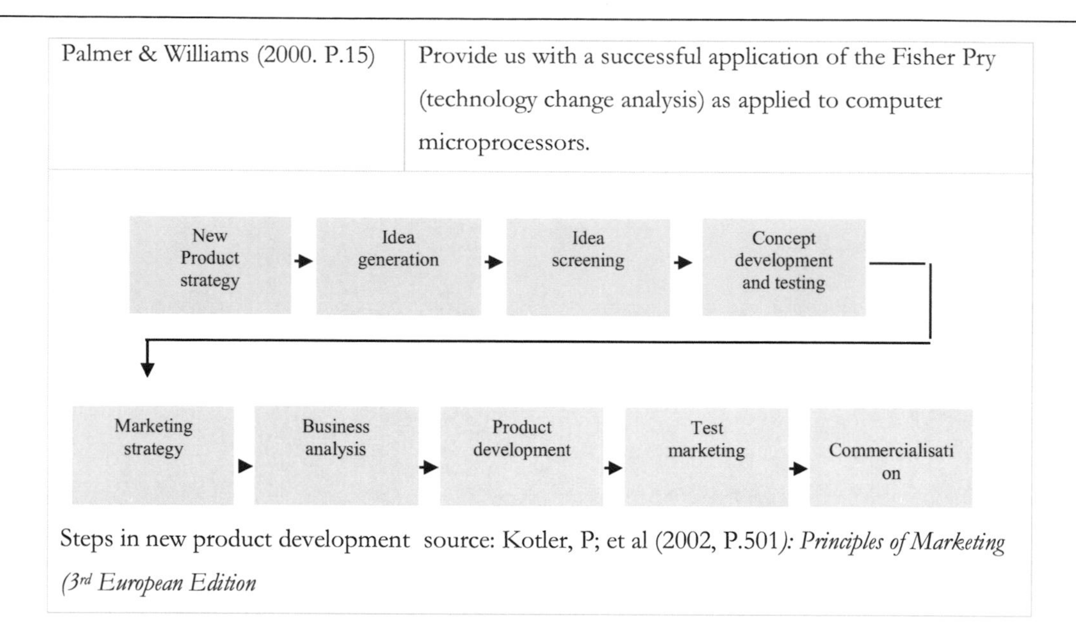

Steps in new product development source: Kotler, P; et al (2002, P.501): *Principles of Marketing (3rd European Edition*

Table 8.38

At this time we will not concern ourselves with the management of this process. This is given within chapter 9. We will concentrate on estimating the material costs and labour costs that are needed to complete this process. This we will do using data presentation in table format (Table 8.39).

Process activity	Material	Labour (See chapter 7.0)	Total
New Product Strategy	N/A	Completed	N/A
Idea Generation	N/A	1k	1k
Idea screening	N/A	1k	1k
Concept development and testing. (separate list is required that identifies material components)	100k	20k	120k
Marketing strategy	N/A	N/A	N/A
Business analysis	N/A	N/A	N/A
Product development (separate list is required that identifies material components)	60k	15k	75k
Test marketing	N/A	5k	5k
Commercialisation (separate list is			

required that identifies material components)*	60k	30k	90k
Totals	220k	72k (3260hrs)	292k

Table 8.39* Denotes that R&D costs must be separated from any specific manufacturing and test plant costs that will be required to produce the products.

We now have our R&D costs that will be analysed by way of their impact upon the P&L for the directional policy employed for our *business plan.* Although presentable, the table given is wrong in that we have not disseminated the data in terms of operational and long term objective analysis. Furthermore if equipment is to be purchased when will it be purchased and must we offset our costs for taxation purposes? It follows that we should provide several tables if we are to be more accountable.
This is important not just for the P&L analysis but also the needs arising out of performance measurement associated with the implementation programme (chapter 9.0) and our performance review and evaluation (chapter 10.0)

R&D will be seen to be a process within a process. That is it has the potential to interact with the sales process. Let us not forget that it is the sales process that will be the prominent factor in the finance of the R&D process which would suggest it will have priority of the engineering hours expended by IVS. This would suggest that R&D analysis be undertaken in a manner where it includes who will be responsible for a given activity including appropriate timescales. All of these factors must be observed when undertaking R&D analysis.

This approach it is hoped will provide more credence to our *how will we get there?* strategy employed. This being said we will assume for this example that we have revised our R&D submittal by way of the following values.

Operational R&D Costs		
Labour	Material	Total
20k	10k	30k
Depreciation factor = 5yrs		
Revised total		22k

Table 8.40

Long term R&D Costs		
Labour	Material	Total
60k	215k	275k
Depreciation factor = 5yrs		
Revised total		103k

Table 8.41 Note that prices have been increased in relation to the previous table to reflect inflation.

It will be seen from the example provided that no additional engineering staff or otherwise is anticipated in order to undertake the R&D activity. Ergo, there is no impact upon the indirect overhead cost.

Packing Costs

Technology has had an impact upon the packing techniques used across a mix of industries. Not least IVS who use the packing techniques employed as a differentiating factor. To this end, a more robust solution is provided without any compromise on the weather sealing properties employed. This they believe is of great importance within their industry owing to a preponderance of unsupervised storage conditions that exist on construction sites.
Patrick's department has produced a formal specification for use by IVS to crate the goods as well as details for handling for the customer base.

The cost of the wood to crate the goods together with all of the sealing materials is common to all of the products. These are included within the material product cost already discussed on an amortised basis. Similarly the special machinery needed to complete the process and labour associated with it, are included on amortised basis within the labour costs already discussed. So in what way will packing impact upon indirect overheads?

Sales process needs are to be considered as packing requires purchasing time. However, this would be classed as one of the functions forming part of the purchasing procedures employed.
For this example we need to concern ourselves with the additional factory/warehouse space arising out of the need to house the packing materials and the packing machine itself. This being said, prudence would suggest such costing factors be acknowledged as forming part of the overall factory leasing cost identified as "any other direct cost" given below. Thus for this example no overhead cost recovery is required.

For the purpose of *business plan* inclusion it is suggested that a paragraph be produced to define the packing philosophy employed by the company.

Delivery Costs

Product freight is known to be a large industry within itself on a world wide basis. To all intents and purposes it is the conduit that is used to link together all of micro environments within a macro environment.

For our example we will need to consider freight industry boundary definition criteria and supply logistics. Boundary definition is a relatively undemanding activity by way of its understanding. World wide documentation is provided to simplify this understanding using INCOTERMS. This document allows purchasers to specify exactly their preference in the form of an acronym or abbreviated term in a way that is fully understood by prospective purchasers and vendors. IVS as a standard operate within an industry where delivery is based upon CIF named destination including offloading on a prepared surface. This term defines that delivery will be undertaken on a *carriage, insurance and freight* basis delivered to the vendors own specified depot. Put another way all responsibility for delivering the product rests with IVS.

Similarly payment terms do not become effective until this specific requirement is completed. The only risks where boundary definition may be unclear with this method of delivery identification is the boundary criteria associated with "a prepared surface" It could be that humid or similar inclement conditions may prevail which could damage the product even though it is packed effectively.

To fully assess any potential delivery overhead cost it is suggested that one defines at which part of the sales process the delivery process begins. It could be pre or post product test? For this example we will consider post test that is, the product is packed awaiting collection from the warehouse main exit. Furthermore we will assume that no special crane or lofying facilities are required to simply transportation within the factory and warehouse environment.

The management of the delivery process is undertaken by the contracts engineers who work for Manfred in conjunction with the Industrial Engineers employed by Hayley. They do not employ a transportation manager or clerk for this process activity. The company employ a firm of freight forwarders who collect the products using trucks fitted with hoisting facilities. These same forwarders complete the remaining part of the delivery process. They also are responsible for providing proof of delivery and acceptance of delivery documentation; important as proof of delivery for invoicing and payment clarification.

Logistically, the delivery process consists of procedures to be followed that define client, IVS and freight forwarder responsibilities. These procedures are such that IVS operate with a mix of three different forwarders. The selection process for these three vendors was based upon cost of trucking facilities and their mileage rate. These are assessed for each contract and included within the tendered price as a direct cost. The reason for this stems from the distance variable across segment geographical locations.

Furthermore the sales process would suggest that delivery costs be identified separately for each tender to allow product invoicing in the event that delivery cannot be completed for matters outside of IVS control.
In summarising the above we will observe that the comments provided refer in the main to the operational business philosophy employed. To all intents and purposes the same comments could be used to represent long term objectives.

However let us not forget that IVS competitors for the domestic segment were shown to provide an as delivered price. This would suggest no major change to this same philosophy other than a modified contract agreement with the three vendors. The relevant delivery costs would then be amortised across the domestic segment product mix costs.

For the purpose of *business plan* inclusion it is suggested that a paragraph be produced to define the packing philosophy employed by the company.

Maintenance

Although not as crucial perhaps in respect of intended expenditure and its impact upon the P&L, maintenance expenditure does have a high degree of importance associated with expenditure identification. The reason for this is the yearly budget allocation and the preparation of a set of chartered accounts that will simplify the expenditure approval process. An analogy could be a request to spend 5k on boiler repairs at the end of the financial year only to find that this work had never been anticipated.
This would suggest that we follow a similar process to that undertaken by the engineering department, namely we should assist the maintenance department to provide an accurate list of maintenance expenditure.

It will be seen that the maintenance department consists of two employees namely one full time electrician and one full time fitter. Contractors are employed to undertake preventative maintenance for factory machinery and all of the cleaning is undertaken using outside contractors.
Additional preventative maintenance involves changing certain electrical and mechanical components but this work is done in-house.

Operational			
Maintenance Purchase Needs	Sales Process Activity	Marketing Process Activity	Cost
Factory & Office lighting & small power replacement	Building regulatory and PLC requirements	Will support promotions policy	10k
Heating & water supply replacement	Building regulatory and PLC requirements	Will support promotions policy	15k

Private maintenance contracts – factory m/c's	Retain machinery availability rates	Will support promotions policy	10k
Building interior & exterior	Building regulatory and PLC requirements	Will support promotions policy	10k
Cleaning contractors	Health & safety requirements	Will support promotions policy	5k
Training		New machinery associated with NPD will prompt training	2k
Total			52k

Table 8.42

Long term			
Maintenance Purchase Needs	Sales Process Activity	Marketing Process Activity	Cost
Factory & Office lighting & small power replacement	Building regulatory and PLC requirements	Will support promotions policy	12k
Heating & water supply replacement	Building regulatory and PLC requirements	Will support promotions policy	10k
Private maintenance contracts – factory m/c's	Retain machinery availability rates	Will support promotions policy	12k
Building interior & exterior	Building regulatory and PLC requirements	Will support promotions policy	15k
Cleaning contractors	Health & safety requirements	Will support promotions policy	8k
New office carpets	Improved ergonomics	Improved ergonomics	10k
Re paint factory	Improved ergonomics	Improved ergonomics	5k
Training		New machinery associated with NPD will prompt training	2k
Total			74k

Table 8.43

Company cars

IVS fully appreciate the importance of competence skills and its impact upon sustaining and growing business activity. They select individuals who are either known to excel in their particular discipline or will establish their selection criteria as part of the interview process. The net result is to reward their employees with a remuneration package that aligns with their status.

All of the managers as well as the contracts engineers are provided with a company car. The latter two individuals need this owing to site travel requirements.
There is also an additional car and light delivery vehicle that is used by the product installation team.

For this example we will suggest that IVS use an outside company car service provider. Here the provider will undertake car lease requirements, car taxation and maintenance and tyre/windscreen replacement. IVS will pay the provider a lump sum based upon the number of vehicles supplied. In addition to this IVS will be responsible for reimbursing the cost of fuel used for business purposes. Thus we can suggest the following simple analysis:

Operational			
Department	Car Lease	Mileage	Total
Management	30k	12k	42k
Engineers	20k	12k	32k
New sales exec	5k	3k	8k
Totals	55k	27k	82k

Table 8.44

Long term			
Department	Car Lease	Mileage	Total
Management	35k	15k	50k
Engineers	22k	14k	36k
Additional Engineers	22k	14k	36k
New sales exec	6k	4k	10k
New sales exec	6k	4k	10k
Totals	91k	51k	142k

Table 8.45

For this example a comment would be included where it is stated that the lease scheme is managed by Eugene's PA. Patrick has the responsibility for all expense documentation and must co sign any expense claim.

Fixed assets

For this example we will state that Patrick has been with the company for two years. In his first year he was given the task of implementing an asset register for existing plant and for new plant. These are all of the assets required by IVS to support the sales process in terms of direct and indirect assets. Here we will use the same analogy as given within the main chapter notes to define how they impact upon the business. Namely, what does the production process need to function and what can be excluded to retain its function?

The process of definition of the asset base is somewhat complicated in that we are reliant upon knowledge of the taxation rules at any instant in time. It follows that variable and boundary definition criteria are subject to change. Even our currently useful guidelines (CUG) techniques cannot be used here for text presentation purposes as that stated could be out of date within months. It is for this reason that our example does not define individual assets such as computers and machinery and descriptions associated with their needs.

Operational & Long Term – Sales Process Asset Base					
Department	Current	Year 1	Year 2	Year 3	Year 4
Executive Suite	2k	2k	2k	3k	3k
Sales	6k	6k	5k	5k	5k
Production	400k	400k	20k*	600k	600k
Accounting	5k	5k	5k	5k	5k
Totals	438k	448k	27k	658k	658k

Table 8.46 * denotes expended existing factory machinery

Operational & Long Term – Marketing Process Asset Base					
Department	Current	Year 1	Year 2	Year 3	Year 4
Executive Suite	-	-	-	-	-
Sales	-	-	-	-	-
Production	10k	10k	10k	20k	20k
Accounting	-	-	-	-	-
Totals	15k	15k	15k	30k	25k

Table 8.47

The values indicated within these tables will not be shown in our P&L tabulation provided later. This is because the asset register is used as part of the P&L that is produced to account for the business at the end of a given financial year. Our *business plan* P&L is structured to account for its impact upon our bank account and cash flow and as such the total cost of the asset will be included.
One of the main advantages of this type of presentation for a technology based organisation is that it draws attention to the cost of assets for each sales and marketing process activity and the impact it has on the margin generated by the P&L. There are other advantages that we can consider as having an indirect impact. For instance certain technologies attract government development funding as a finance initiative. This method of presentation will allow a more accurate assessment of the amount of funds needed. That is how much will it cost to produce and how much will it cost to commercialise?

We could also suggest that any sales undertaken by IVS will have an impact upon the client's asset base register. Here the author would recall a NPD/innovation opportunity to increase the energy loading of overhead power lines. The development was stopped owing to the cost of a special camera to measure temperature at a cost of 30K. The development scientists thought this too intrusive for clients and as a consequence terminated the development. It later transpired that the camera if used correctly would have saved the cost of 110 pylons at a cost of 130k each however, for this opportunity the initiative had been lost.
Yes the analogy provided would have been assessed differently using a marketing plan but it does draw one's attention to the tertiary impact of an asset base for a vendor and a purchaser.

To simplify the example put forward any reference to tangible assets is excluded. That is those assets that could be readily exchanged for cash. The company accountant or company secretary should be approached where any such specific information is required as part of the *business plan.* Note that the IVS factory operates using leased facilities and thus land and buildings do not form part of the asset register.

Professional services

We have already observed that HR, legal and specific advertising material is sub contracted to an outside resource base. All of the costs associated with these companies impact upon the business as a whole and as such must be included as an indirect overhead cost. Before we begin to analyse our cost base for these items we must provide comments on their need from an organisational point of view. This is made easier to understand if we revert back to the need for employing our marketing plan and the limitations imposed by that undertaken to arrive at this part of the *business plan.*
Our marketing plan will maximise our opportunities and limit our risks but here the latter relates more to our product or service strategic intent. Risks will be seen to abound across the business micro environment that is caused by macro environmental factors and vice versa either by way of use of product or service and/or regulatory factors. Thus our comments must be specific in that how is it possible to reduce the risks imposed by these factors? Sub chapters are used to analyse each outside resource.

Human resource or personnel needs are managed by Eugene's PA. The documentation that is in force to employ individuals and to manage their employment is available in standard documentation format. The resource provider visits the company once every month to review any ongoing issues that may result in dispute. The same provider also offers a 24 hrs hotline for any potential emergencies. Organisational manning levels are such that they do not warrant a full time personnel manager.
Furthermore, the cost of the service as given below would warrant using the service provider. It is to be noted that any change of PA status may require a review of company procedures.

The legal resource provider consists of a local firm of attorneys who specialise in contract law. They were employed at the outset of IVS to create standard terms and conditions for sale and tender for product and service supply and installation. These standard documents form an integral part of IVS tender documentation and contract handling procedures.

Where any disparity occurs the service provider will review at an extra cost and support the company accordingly. The marketing plan employed as part of this *business plan* would suggest that we review the legal procedures undertaken by the business.

Corporate intent is seen to increase IVS services in respect of market attractiveness. Currently, all of our target market terms and conditions are rejected in favour of IVS standard documentation usage. Mainly because the segment market intermediaries and end users use conditions that are generic in format that apply to all product or service types.

This procedure provides a more purposeful adaptation of law and associated remedies in relation to IVS product mix and its intended usage. However, the change in IVS product mix and the NPD work would require an update of existing documentation in particular any risks that may impact upon the macro environment.
Similarly, strategic intent would suggest that segment documentation be reviewed at the tender stage to create more purposeful application (aligns IVS & client needs) during the contract handling and warranty process. The overall impact will be to position IVS within the market segment not just in technological competence but also business competence.

Any analysis of enquiry documentation would suggest the employment of a full time legally trained individual who has the competence to adapt IVS current procedures to strategic intent. The relevant impact upon the P&L is given below, Tables 8.48 & 8.49 refer.

Advertising costs are shown to consist of a reduction from operational to long term requirements. This is because of the reduction in market needs arising out of the transfer of product and company brochures to the company web site. Furthermore, for those market segments who still insist upon brochures in hard copy format then the new printers to be purchased are known to produce quality documents.

Operational Resource Provider Activity	Cost
Human resources	15k
Legal resource – Yearly fee	15k
Legal resource – Consultancy fees	10k
Advertising material	5k
Total	45k

Table 8.48

Long Term Resource Provider Activity	Cost
Human resources	17k
Legal resource – Yearly fee*	17k
Legal resource – Consultancy fees*	15k
Advertising material	Nil
Total	49k

Table 8.49

* Denotes the *business plan* would suggest that these two services could be undertaken in-house without any significant impact upon the margin generated.

Insurance Costs

Chronologically one would suggest that this particular indirect overhead cost analysis would always follow any sub chapter consisting of any legal impact upon the *business plan.* The main reason for this is its direct association with risk limitation factors. To this end what risks are eliminated by virtue of the irrefutable paragraphs contained within the company's terms and conditions and those that are not? Similarly, in what way is IVS legally obliged to provide employee liability and other regulator insurance needs?
We must also consider the macro environmental factors alluded to in respect of any consequential losses and the limits of liability.

It is unlikely that a given individual within an organisation will be able to fully answer the needs of this sub chapter. Rather it is suggested that the existing insurance provider be contacted. The implications associated with the factors described within the opening chapter should be discussed in relation to summarising and confirming the status of existing insurance cover and that which will become relevant for long term objective requirements.

Normally stated as a percentage of total sales the insurance premiums should have a monetary value for insertion within the P&L. This is given in simplistic format; Table 8.50 refers:

Premium Basis	Operational Cost	Long Term Cost
Regulatory premiums	37.5k	55k
Product/Service related premiums	12k	15k
Total	49.5k	70k

Table 8.50

Care must be taken to identify risk limitations that do not apply to specific individual client requirements. These can always be assessed separately and its premium included as a cost for that particularly contract. Hence the insurance costs should be separated as an individual entity to the external resource base given within the previous sub chapter.

Stationary costs

Possibly the most understood potential indirect cost for consideration by any company or is it? All company departments use stationary but which departments use the most and do the needs of different departments create different stationary ordering needs? Here the latter is unlikely to occur in that one will probable use that which is available. Indeed, how many of us have wondered what to say to the boss's secretary on a given day in order to obtain a new pen? This in turn gives rise to an argument to reassess interdepartmental stationary strategy.
Similarly, is the company known by way of its use of stationary? It could be the letterhead employed or the business cards presented by company staff. Furthermore is there any strategy associated with font type, paragraph spacing or a specific colour of paper for a given reason. Indeed, the process used to define all of these factors could digress into other areas of the business. Different coloured telephones for use with customer lines? All of which would support the need for a stationary strategy.

Sales are the most probable high intensity stationary users. More importantly professional presentation would suggest they use paper of a much better quality than that used for filing purposes. Tender documentation is also a crucial factor in the initial assessment of a formal submission. The type of folder used for instance and the methods employed to identify and divide subject matter. Clearly, the levels to which one would aspire, is to a large extent dependent upon client aspirations and that undertaken by the competitors.

For this example therefore we will accept that a formal strategy has been produced the details of which we will assume are included within this sub chapter. Our indirect cost analysis would follow in a simple format as shown in Table 8.51.

Stationary Costs	Operational Cost	Long Term Cost
Office (define by dept)	15k	17k
Manufacturing	5k	6k
Total	20k	23k

Table 8.51

Advertising

To many this sub chapter is marketing in its entirety. In that these are the costs associated with promoting the company other than company brochures. However, our promotions policy as undertaken within chapter 7.0 would show that advertising or promotions forms an integral part of the strategic intent of the business and the direction it will take.
Our costs associated with the latter will be seen to be very much dependent upon how we will manage the promotions policy. There are basic needs such as lunches and promotional gifts in the form of pens and mouse mats. Indeed, Christmas cards and calendars must also be taken into consideration. However, the market audit would suggest a need to attend more trade shows but which shows must we choose? Clearly some form of further research is needed. A phone call to a set of important clients may be all that is required.
We must also consider to what intent we will promote the new product mix and where? And will we need to manufacture a specific unit just for advertising purposes?
For the purpose of this example we will provide a lump sum price for the basic promotional needs. This being said prudence would always mandate that individuals undertake a thorough assessment of all associated costs and tabulate them accordingly. Let us not forget that our same directional policy promotes more empowerment within the organisation.

It follows that some one of standing within an organisation must produce a *business plan* where budgeted amounts reflect correctly the needs of the marketing plan. It will then be left to others to store this same information in a manner that will allow its approval or otherwise at the time of expense needs. Here it could be argued that senior management may be away from the office and that a nominated individual has been asked to approve the purchase of a promotional item. He or She would need to know how the proposed cost of this item aligns with that budgeted.

Promotional Strategy Costs	Operational Cost	Long Term Cost
Standard promotional items	15k	18k
Trade journal advertising	2k	3k
Trade fair activity (includes lodgings)	45k	65k
Sample micro wave unit	-	2k
Sample solar pump unit	-	500
Total	62k	88.5k

Table 8.52

Training Costs

Training requirements for any company would relate to company interdepartmental needs as well as that of clients arising out of the promotion of the usage of a specific technology. In most cases clients will pay for the training of their staff and as such the appropriate costs should be included within any tendered proposal or quotation for as a specific need.

Client training can become a crucial factor at the time of tender/quotation analysis. If the client is large by way of the number of employees they may have, they may not wish to retrain and would by way of ease of operation pay a premium for a competitive product or service. Such factors it is hoped will have been identified at the time of market audit, SWOT analysis and strategy formulation which in turn would provide for cost assessment within this sub chapter.

For the purpose of this example we will suggest that there is no expense associated with client training. IVS on the other hand will need to undertake some form of training arising out of the new technologies to be employed for the long term objectives. There is the diffusion of this technology within IVS. There is also the purchase of new equipment determined by direct and indirect overhead sales process support requirements.

To all intents and purposes the level of training will always be dependent upon how those who are to be trained will behave (Hooley et al (2004. P.493)). This would suggest that we will always be reliant upon the product or service provider we choose for any new equipment to possess their own marketing analysis tools to analyse our needs.
The latter we could argue could be assessed in the form of "one off" training or ongoing training functional needs. For example certain software packages may be known to require specific updates over a given period. It is possible that an accounting data base has been purchased at a given cost in relation to a stated scope of supply. Additional amounts will be paid over a three year period to increase the functionality of the data base over the corresponding period. What training needs will arise out of this type of purchase?

Here the emphasis is placed upon the need to question all departmental supervisors accordingly to ensure that no extra costs will be borne by the company that did not form part of this *business plan.* Similarly in what way are employees "worked hours" accounted for either as a sales process activity and/or as a business expense?
If all "worked hours" are assessed as a sales process cost then certain costs must be included within our cost analysis. For instance if an engineer amortises his time over a number of contracts and our situation suggests he must undertake four hours of H&S training then we must show that as a training cost. If he or she does not book time then we can disregard his contribution to training costs. That is we must only include the costs associated with the trainer or training programme as well as its venue.
Finally, there may be a relaxation of taxation associated with training of individuals. It follows that the impact upon the business created by training needs should have an additional comment that describes its tax detracting impact.

Training Costs	Operational Cost	Long Term Cost
H&S training	5k	4k
Individual academic support	10k	-
IVS NPD diffusion	2k	5k
Journal research activity	-	4k
New process equipment	1k	2k
Total	18k	15k

Table 8.53

This particular sub chapter warrants careful consideration to the extent of its potential to link via computer software to its impact upon the P&L statement. Training is known to increase the knowledge base and as such should by way of marketing plan usage may or most probably will give rise to a mix of needs arising out of how the "internal factors" impact upon strategy. This strategy must be disseminated by way of sales process activity, indirect overhead activity and *business plan* governance (chapter 10.0). There exist therefore many solutions for consideration hence a simplified method of determining the impact upon the P&L each or a mix of needs may have.

Other indirect costs

Here there is the potential for no extra defining criterion for this example other than listing the appropriate latent costs that will impact upon the IVS *business plan.*

Indirect Cost Factor	Operational Cost	Long Term Cost
Land & Buildings*	95k	130k
Facility transfer costs*	-	15K
Office Furniture	5k	10k
Printing machines	3k	-
Flowers	1k	2k
Gardener	1k	2k
Office cleaning	2k	3k
Window cleaning	2k	3k
Fire extinguishers	1k	2k
Security systems	1k	2k
Fork lift maintenance	2k	4k
Total	113k	173k

Table 8.54

Land and buildings as stated previously are leased from the local municipality. In reality significantly more information would need to be produced for this particular *business plan.* To this end, long term objectives have identified a need for larger premises. On what basis has someone arrived at the cost of the larger premises?

Factory floor and office utilisation needs must be identified as they relate to current process activity as well as that projected for long term requirements. Clearly, for this example a leased facility could suggest that IVS have had to adapt their business activity to generic building availability. What impact upon the sales process would a purpose facility have and what would be the cost of such a facility?

While any decision arising out of any change in this situation will rest with business owners or senior management team they will need factual information to decide and act accordingly. For this example therefore we could argue that a separate overhead factor be produced in the form of "Facilities overhead" cost. This could be structured using a formal SWOT analysis and to then observe its impact upon strategy formulation and business direction needs.

We can recall from our market audit that IVS already has the most impressive facility within the business macro environment. However, their competitors are already known to exist much closer to the geographical location of the industrial segment market intermediaries.

Furthermore, any analysis it could be argued should take into consideration the ergonomics associated with IVS employees and the clients who visit the facility. Here the author would recall a question put to a prospective client back in the 1980's when shown a photograph of competitor equipment. "Why is the factory floor painted green"? was the question asked. The answer given was "It's not paint its carpeting". Here the use of carpeting showed all that viewed it that a high level of cleanliness prevailed in a manufacturing environment where any dirt was detrimental to product quality.

This final overhead cost factor analysis would further support the need to undertake an iterative process when undertaking the preparation of an effective *business plan.* Furthermore we could use the same method of study to state that *business plan* content is not generic in nature but is reliant upon the use of a formal marketing plan. It follows that the selection process used to define and react to the latter will have a direct impact upon the strategic aims of a *business plan* as suggested in chapter 1.0 and 2.0 hereof.

We can now begin to examine how the foundations of the chapter analysis undertaken influence a formal P&L statement, Table 8.56 overleaf refers.

Operational

Total Sales	15,011k
Total Manufacturing Cost	7.056k
Gross margin	7,955k

Cost of Staff	2,195k
Cost of Utilities	5,28k
Engineering and Design (E&D)	30k
Research and Development	22k
Maintenance	52k
Company Cars	82k
Fixed assets	453k
Professional Services	45k
Insurance costs	49.5k
Stationary Costs	20k
Advertising	62k
Training	18k
Facilities	113k
Projected bonus payments	1,501k
Net Margin	**3307.22k**

Table 8.56

Normal practise would suggest that the P&L should have several right hand monetary columns that relate to the following years forming our long term objectives.

The impact this would have upon this example would be one of potential confusion owing to the mix of costs for each year. Our example has projected the relevant costs for three years hence, that is *where are we now* and the costs of "where do we want to be" as an operational P&L and *how are we going to get there* as our long term P&L analysis.

Long term

Total Sales	22,720k
Total Manufacturing Cost	10,150k
Gross margin	12,570k

Cost of Staff	2,986k
Cost of Utilities	8,3k
Engineering and Design (E&D)	32k
Research and Development	103k
Maintenance	74k
Company Cars	142k
Fixed assets	2,688k
Professional Services	49k
Insurance costs	70k
Stationary Costs	23k
Advertising	88,5k
Training	15k
Facilities	173k
Projected bonus payments	2,272k
Net Margin	**3,846k**

Table 8.57

Many statistics can now be used to define the status of our projected P&L. The most important statistic is the percentage profit to sales which is obtained using the ratio of net margin to total sales. Thus for this P&L our value is 22% operational and 17% long term.

This ratio would suggest that we have increased our sales by 51% yet our increase in net profit from a figurative amount is only 16%. While this may prove unacceptable to a select audience we can argue that the P&L reflects competitive market status. How would our P&L appear without the use of the marketing plan (*Existing product/Existing market Vs Market Penetration*)?

Clearly, the tables provided exclude any projected monetary amounts for taxation and/or shareholder premiums. The company accountant or company secretary will need to complete any specific assessment that may be required.

Another important statistic is that which allows us to account for the cost of overheads using sales process performance measurement. Most companies will use the labour expended as part of the manufacturing process to account for financing the cost of overheads. This will allow a direct correlation and congruency with the cost of producing a sale.

Put simply if we deduct the net margin from the gross margin and then divide this by the number of hours we will define our overhead rates. This is given as (or similarly presented):-

Operational overhead rate

$$= \frac{\text{gross margin} - \text{net margin}}{144\text{khrs}}$$

$$= \frac{7955\text{k} - 3307.22\text{k}}{144\text{k}}$$

= 32.28/hr (21.53/hr less projected bonus)

= 322.8% based upon £10.00/hr factory rate

= 215.3% (less projected bonus)

Long term overhead rate

$$= \frac{\text{gross margin} - \text{net margin}}{162\text{khrs}}$$

$$= \frac{12570\text{k} - 3846\text{k}}{162\text{k}}$$

= 53.85/hr (39.83/hr less projected bonus)

= 489.5% based upon £11.00/hr factory rate

= 362% (less projected bonus)

Thus our hourly rate for sales purposes are 32.28/hr (plus 22% - divide by 0.78) and 53.85/hr (plus 17% - divide by 0.83) for operation and long term projections respectively.

There are no specific boundary limits we can associate with this statement. However, suffice it to say if we do not achieve our projected sales targets then we will expend less manufacturing hours which in turn will create what is termed as "under recovery".
In other words we are not generating enough hours to meet our expenses. It should be noted that the long term projected overhead rate could be considered high; albeit this is reduced by deducting the performance related bonus scheme payments.

The major factors that impact upon this high rate are salaries and the cost of new manufacturing plant listed as a new asset. In the event that the recovery rate is unacceptable we must revisit the cost of this new plant. Could it be leased? Can we buy just the one machine and operate using more than one shift pattern? Regrettably only a trial and error exercise will provide an end to our aims in this regard. The same it must be said for any other factor that constitutes our overhead cost base.

Before we have reached this particularly part of our *business plan* the company accountant will most probably be already seeking projected cash flow for the business activity. The business it can be seen operates in a high value industry that attracts a large volume of costs. Based on our projections these are shown in Table 8.58 & 8.59.

	Operational*	
Cost Base	Total p.a.	Monthly Average
Manufacturing	7,056k	588k
Direct & Indirect O/Hs	3,147k	262.25k
Total	10,203k	850,25k

Table 8.58

	Long term*	
Cost Base	Total p.a.	Monthly Average
Manufacturing	10,150k	845.6k
Direct & Indirect O/Hs	6,452k	537.7k
Total	16,602k	1.383.3k

Table 8.59 * denotes bonus allocation excluded

Therefore in what way will our projected sales output on a monthly basis exceed the values given? Put simply what are our cash flow projections? The onus for providing the appropriate data will always rest with the sales department. For this example we will simplify the calculation by dividing the total sales by 12 months to arrive at our projections.

	May				June			
Product/Service Type	Qty	Total Sale	Total Cost	Margin	Qty	Total Sale	Total Cost	Margin
A - Steel	1	100k	39.24k	60.76k	1	100k	39.24k	60.76k
A - F&B	2	204k	78.48k	125,52k	2	204k	78.48k	125,52k
A - P&P	1	95k	39.24k	55.76k	1	95k	39.24k	55.76k
A - PC	2	240k	78.48k	161,52k	2	240k	78.48k	161,52k
B - Steel	1	62k	34.91k	27.09k	1	62k	34.91k	27.09k
B - F&B	1	64k	34.91k	29.09k	1	64k	34.91k	29.09k
B - P&P	0	0	0	0	0	0	0	0
B - PC	2	280k	139.64k	140.36k	2	280k	139.64k	140.36k
C - Steel	1	54k	29.13k	24.87k	1	54k	29.13k	24.87k
C - F&B	0	0	0	0	0	0	0	0
C - P&P	0	0	0	0	0	0	0	0
C - PC	2	240k	117,24k	122,76k	2	240k	117,24k	122,76k
A	2	160k	75,24k	84,76k	2	160k	75,24k	84,76k
B	1	35k	18.73k	16,27k	1	35k	18.73k	16,27k
C	1	30k	13.92k	16.08k	1	30k	13.92k	16.08k
A	25	25k	16,75k	8,25k	25	25k	16,75k	8,25k
B	100	6,25k	4,364k	1,886k	100	6,25k	4,364k	1,886k
C	25	12,5k	9k	3,5k	25	12,5k	9k	3,5k
Sub Total		**1,607.8k**	**729,27k**	**878,48k**		**1,607.8k**	**729,27k**	**878,48k**
O/H costs				262,25k				262,25k
C/F Sales		1,607,8k		616,23		3,215.6k		616,23
Net Margin				616,23				616,23
Accrued Margin		From previous month		1000k				1,616.23 k
Net balance				1,616.23k				2232.46

Table 8.60

For simplification purposes only two months have been inserted within the table given. Any presentation material must however include the total twelve month financial period. Furthermore

given that the values stated are projected, one would suggest that any cash flow projections be omitted for long term business objectives unless specifically mandated.
What must be borne in mind is the operational importance this particular table has on business operations, namely sales, manufacturing and purchasing objective criteria.
We can conclude our chapter example by examining how we have achieved the aims of the four bullet points associated with Fig 36 contained within the main chapter.

Competence Resource Base	Comment
Productivity Vs Value/Volume	The structure of the P&L statement produced as an integral part of the marketing plan example can be seen to allow an assessment of each applicable factor and its impact upon value or volume. What it does not state is the process or any algorithm that may be employed to provide a further evaluation of its strategic impact upon the evaluation of productivity. Where such a process or algorithm is available or analogous research supporting documentation then its usage should be referred to as part of the *business plan* presentation material used. The structure of the *business plan* employed would suggest a useful point of origin from which any further research on productivity could be undertaken.
Existing product/Existing market Vs Market penetration	It can be seen that competitive forces will always have a major impact upon this bullet point has upon the P&L statement. Fortunately significant primary and secondary data is available that will allow further research if so required. However, it can also be observed that micro environmental business ego's will allow detracting factors in respect of macro environmental needs to be observed in relation to the P&L statement. Regrettably limited text book data and research journal is available that considers such factors (Meldrum (1994)).
Existing product/Existing market Vs New products/New markets	The impact this bullet point has upon a P&L statement is the difference one would observe from that which is structured using operational needs and that which considers long term objectives. It will also be seen that this particular bullet point and that which it follows will have a direct relationship on the *business plan*

	implementation programme discussed within the next chapter. To this end, neither will impact upon the business management techniques employed but will most certainly impact upon the management techniques employed for the marketing plan. The potential this creates for significant more evaluation of management philosophy is worthy of further research.
Value of new product Vs Development for new markets	One of the draw backs we have observed with this bullet point is the crucial impact upon market intelligence, SWOT analysis and the strategies to be adopted from the facts observed. Indeed, it can be similarly observed that we have omitted from this chapter example any reference to the "integrated refrigerant" product. This it is argued requires further analysis of the implementation techniques employed (chapter 9.0) and the measurement of performance (chapter 10.0) for those companies who exist currently without a formal marketing and *business plan.*

Table 8.61

Thus we can complete this section by creating a sheet to start our next chapter.

Fig 38
Strategic Analysis – How are we going to get there?

The Marketing Plan Planning Process"

"Business Plan Implementation Programme"

Page 1 of 1

The remaining portion of this chapter is used to provide supporting text for marketing strategy within a technological environment.

Supporting business plan research criteria

Most if not all of the textbook data on this subject provide general statements that define that budgeting must be done in relation to resource allocation. Piercy (2003. p.579) observes that limited data is available on the subject and has endeavoured to suggest methods that allow a more reasoned approach to the subject as a whole.
The importance he puts on the subject is summed up in terms of the following statement:

> "There can be surely few things more futile and damaging than to get people excited about new plans and innovative marketing strategies and then refuse to resource these plans and strategies".

This implies that there may be a gap between what management wanted and the cost of what marketing is proposing and as such "contingencies" must be borne in mind from the outset of the market planning process – Piercy (2003. p.580). Here we can transpose "the *business plan*" for "marketing" and arrive at an argument that supports the use of marketing as a science for the basis of our *business plan*. To this end, the marketing plan put forward hereof provides for all of the questions we must ask ourselves that must be used to create the latter and how to structure our necessary response criteria.

It is argued that the adaptation of a marketing plan in respect of its impact upon a company *business plan* would offset many if not all of the arguments put forward (Piercy). To this end, we are not presenting a formal marketing plan using singular marketing procedures. We are presenting a *business plan* where it can be seen that the marketing plan selection criteria and its usage impacts and supports immensely the chronological process employed to arrive at a formulated *business plan* and P&L analysis.

Finally there would exist a further argument in respect of the indented quote given above by way of the need to further define the use of "resource". Yes this may refer to supporting financial resource however new millennia observations would suggest we expand "resource" to include academic and skills availability.

Implementation Programme

So far we have reviewed the process associated with the preparation of a *business plan* using a typical marketing plan as our road map. Those of you who have a wish to implement such a plan or similar plan would wish to heed text book author's assertions in this regard.

> "Any manager or organisation which is not familiar with the concepts and ideas discussed this far is bound to be at an immediate competitive disadvantage compared with someone who is"
>
> Source: Baker (2000. p. 468) *Marketing Strategy and Management (3rd Edition).*

The manner in which it is implemented therefore must have an impact upon the business as a whole. As a rule of thumb a successful *business plan* using marketing to undertake its strategic objectives and planning strategies will provide the basis for a company's strength and its revenue generating activities.

The manner in which it is implemented therefore must have an impact upon the business as a whole. Here it is interesting to observe how text book authors describe the importance of the implementation programme.

> "One firm can have exactly the same strategy as another yet win in the market place through faster or better execution".
>
> Source: Kotler Et Al (1997. p. 104) *Principles of Marketing (3rd European Edition).*

The main purpose of having an implementation programme allows a manager or managers or executives and owners to exercise control over an organisation or a business and its activities and so give it direction and purpose. Piercy N. F (2003) describes the types of management techniques that could be employed as adapted in Table 9A.

Management Performance	Non Visionary Management	Visionary Management
View of the future compared to that of the competitors	Conventional and reactive	Distinctive and far sighted
What absorbs more senior management time?	Reengineering core processes	Regenerating core strategies
How do competitors view your company?	Mostly as a rule follower	Mostly as a rule maker
What is your greatest strength?	Operational efficiency	Innovation and growth
What is the focus of your company to build competitive advantage?	Mostly catching up	Mostly getting out in front
What if anything has set your agenda for change?	Your competitors	Your foresight

Do you spend most of your time monitoring the status quo (the maintenance engineer) or designing for the future (the strategic architect)	Mostly as an engineer	Mostly as an architect

Table 9A: Source, adapted from Gary Hamel and C.K. Prahalad (1994) Competing for the Future, *Harvard Business Review*: July/August 122 - 28

However, no specific set process is available within text book format that accurately describes in a simple format how to implement a marketing plan for a specific business activity. To a certain extent marketing as a science would suggest the use of Internal Marketing to achieve an end to a required marketing aim.

Put simply, it is expected that the use of internal marketing will, at the very least prompt most company individuals or at least the management team to have a higher degree of perception, acknowledgement and the use of suitable techniques to rectify any market turbulence or lack of profit earned through their actions. Or simply to improve upon what may already be a highly efficient organisation. Hooley et al (2004) suggest that there are various types of internal marketing for consideration.

- "Internal marketing that focuses on the development and delivery of high standards of **service quality** and customer satisfaction;

- Internal marketing that is concerned primarily with the development of **internal communications programmes** to provide employees with information to win their support;

- Internal marketing that is used as a systematic approach to manage the **adoption of innovations** within an organisation;

- Internal marketing concerned with providing products and services to users **inside the organisation**; and

- Internal marketing as the **implementation strategy** for any marketing plan".

These facts notwithstanding for some organisations where marketing graduates are not employed, the implementation of marketing throughout the company environment may be seen as too much of a radical change.
This does not mean that the use of certain marketing techniques be neglected, especially if they are proven methods of company enhancement. Here we can observe certain aspects of marketing to further expand our knowledge base but to also understand why this increase in knowledge is important to the implementation of the *business plan.*

Detracting factors for any effective use of knowledge may be the barriers created by specific company departmental supervisors or managers. This could be borne out of having no commercial acumen that will allow a given individual to understand the relevance and importance of that put to them. One such department is arguably an engineering based activity. Indeed, the interface between R&D and marketing is very much a well discussed research journal topic. Rarely if any do we see research in respect of the interface between marketing and sales or accounting or manufacturing.

The reason for this could be that all company departments other than engineering or R&D can be considered as a "receiver" and "responder" departments whereas engineering/R&D could be assessed as a "provider" department. Here the latter creates a specific technological solution using their competence but is then reliant upon others to secure orders, manufacture, deliver and account for the business activity.

This method of description it is suggested is closely aligned with "Sales" and "Marketing" company activity definition. To this end, a sales based organisation takes what has been built and looks for markets. Marketing based organisations on the other hand endeavour to create a level of fusion that allows environmental factors to drive technology advancement and its usage at any given time in a balanced manner.
This in turn may give rise to further factors of evaluation techniques that must be considered when assessing marketing plan usage.

Here the marketing plan employed throughout this book (Fig 11) can be considered as generic by way of its relevance to business application. There is however a need to create a level of fusion between that which is apparent or indeed real into a balanced objective and consequent strategy. Put simply we must define the boundaries associated with the knowledge base one would use to support the structuring of our market plan and application congruency across the business as a whole.

Fig 39 overleaf is used to describe the latter for the example used throughout that is a technology based company. Other business types such as the pharmaceutical industry or the metals industry for instance would need to change the box referenced "Technology" accordingly to reflect their industry evaluation criteria.

Proof that a marketing led strategy can improve a companies overall performance is borne out using an interesting example (Coates & Robinson (1995)). Moreover, the example is particularly of interest in that it involved a technology based industrial manufacturer that possessed no marketing department whatsoever. The inference being that the marketing practices employed started virtually from scratch but when implemented created a significant impact upon business performance.

9.1 Ontological assessment

Somewhat of an elaborate word for what is essentially an awareness of the market environment within which the company operates within and how marketing by your company or competitors can impact upon this same environment. This is depicted in a simple format as shown in Fig 40 below. The purpose of this model is to identify how your market is structured and how marketing has an importance across the segment as a whole and its association with Fig 40 overleaf.

Its impact upon "marketing plan implementation" is to inform the constituent parts of the market environment of the context of your marketing activities relative to your position within the environment. Suffice it to say that the paradigm shown must be acknowledged in all aspects of the *business plan.*

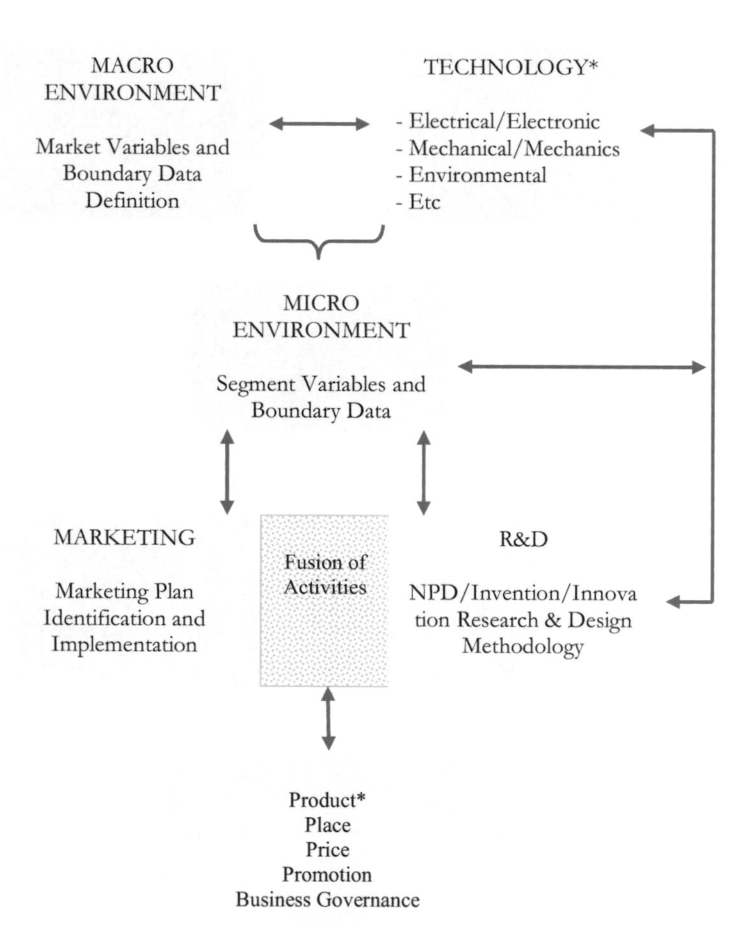

Fig 39 Marketing Plan Knowledge Base Assessment Criteria

* Denotes refer to sub chapter 9.2 below.

For information purposes the environment shown is discussed in a way where the overall environment is identified as the macro environment. The constituent parts that make up this macro environment are referred to in the form of micro environmental factors.

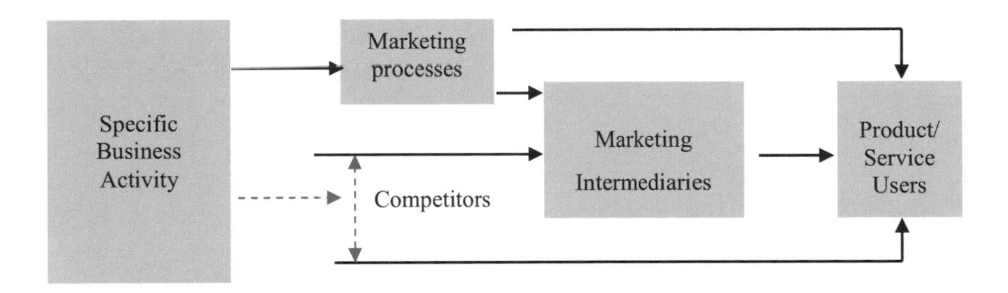

Fig 40 Modern marketing system adapted from Kotler, P; et al (2003).

9.2 Epistemological consideration

Another (and final) elaborate word that has a degree of relevance that is again relatively straight forward in its application. In the first instance we must not forget that most industries are classified as a specific market segment that attracts statistics that are collected and collated by national statistical authorities.

It follows that a marketing based *business plan* will not only help to reinforce and diffuse the factual information already provided for a given industry but have the potential to expand upon the statistics provided. Epistemology is the term used to refer to the study of the source and nature of information as well as any limitations of knowledge.

Its relevance for a marketing plan or indeed a *business plan* impacts not only upon market statistics but also upon every department operating within a business environment. Here we must consider the standards we use associated with all variables and the boundaries that they operate within. Current taxation laws, remedies associated with contract law, existing and planned technologies, speed and safety of manufacturing equipment are typical examples for consideration.

However, we must also consider how all of these and any other variables have the ability to impact upon operational and/or long term strategy in relation to how they evolve.

Research shows that micro processor and other technologies have a linear growth up to a certain point until such time new technological advancements allow them to increment further (Palmer & Williams (2000). During such time a technological advancement occurs, vendors product mix enhancement is shown to incorporate more features which may not incorporate a user benefit but may not form part of a competitors features. Clearly, epistemological data in respect of known or anticipated change in standards should be observed as part of any *business plan* and commented upon accordingly; fig 39 refers to this fact accordingly.

9.3 The Planning Process

Irrespective of the business under review a planning process should be structured using four constituent parts. In model format this is shown in Fig 41 (or preferred similar)

These four constituent parts can be analysed separately in respect of the marketing plan process undertaken previously.

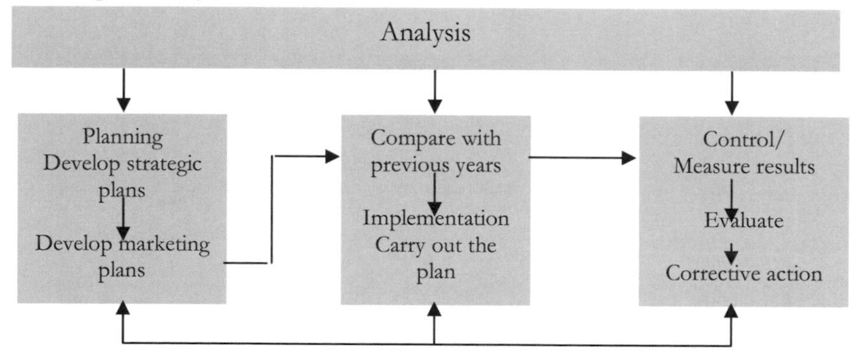

Fig 41 Modified Market Analysis, Planning, Implementation and Control.(Kotler, P et al)

9.3.1 Analysis

Here we need to understand the environmental factors associated with the market in the context of variable definition in respect of the boundaries that they operate within. The market audit together with the SWOT analysis undertaken will have already provided the information required for the analysis phase. In summary, there is a need to have collected data in respect of the bullet points listed.

- What markets exist, how are they segmented?
- Who are your competitors (list for each product type)? What are their strengths and weaknesses, what is their product mix, their competitive position and business direction?
- What is your product portfolio? What are your business strengths in relation to market attractiveness (Fig 27, Page 107)

Where no marketing is employed it may prove interesting to analyse how the business is structured using the bullet points given above. Evidence exists, particularly where companies are structured with a mix of individual business (BU) units, that BU managers may not be familiar with corporate objective and the strategies that must be employed to achieve them. Here the manager has only one interest, to meet the projected sales requirements based on lowest priced solutions. Moreover, what the sales department is doing to identify who will offer low priced offers and on what basis?

Last but not least is a need to analyse the company assets in respect of existing and potential new target markets. A factor it must be said that has the potential to be more important than a combination of all the bullet points given above.

A company's asset base attracts much literature. To all intents and purposes it is the assets that have a major impact upon company value. However, the author would put forward a strong argument that the asset base has the potential to be vastly under estimated.

In what way are the assets used? If a "Sales" based organisation the asset will be a resource. If a "Marketing" based organisation the asset will have a level of competence applied to it in order to respond to the needs of the bullet points given above. An analogy could be a dairy farmer who owns 60 acres of prime land. The land produces the best grazing available the impact of which is that the farmer produces the best milk available with the same impact on profit. It is bought by a larger "sales" focused organisation who then sells it owing to drought conditions.
The purchaser (marketing focused) then begin mining operations as the land has numerous ore deposits.

9.3.2 Planning

Strategic planning is all about deciding which marketing strategy will benefit the strategic objectives set by the business owners or managers.

- Review your objectives in terms of new and existing products, market penetration and provide financial forecasts in respect of turnover.

- Is there a gap analysis in the context of product mix Vs market mix, prices and promotion?

- Do you have a SWOT analysis in terms of the critical success factors the individual importance of these factors and their comparison with competitors?

- Arising out of the SWOT analysis what are the key objectives and key strategies; also the key issues that must be addressed and the financial consequences of them?

- There is a need to consider individual segment strategies and to convert them into specific action plans including responsibility for the action, its timing and any finance requirements.

- Check statistical data that may be available from external sources in terms of market share opportunities, growth markets, prices and customer spending assessment & criteria.

- Does the statistical data support the objectives put forward by the organisation.

There is a requirement to separate these fundamental needs of the marketing strategy from the processes used to implement them. Processes it must be said tend to focus on the individual department that they operate within. They can quite often disregard the information that forms the basic building blocks of departments they feed. De Bono once said that if we are building a garage forecourt using interlocking bricks we cannot expect the overall installation to be exacting if the first brick we lay is wrong.

It follows that the employment of processes within industry have become very much a quality function hence the previous reference to ISO 9000 and its derivatives. However, for those of you already familiar with such processes you may now be aware following the introduction of a marketing plan that they have a tendency to focus on a micro environment. That is, what is it the company must do rather than how does the process impact upon the macro environment.
One major factor that could act as a differentiating factor with a more efficient sales process is that it better informs the customer of your own company activities. There is also the impact upon cash flow to be considered.

9.3.3 Implementation

Here there is a need to create benchmarks that are crucial to the success of the planning cycle. This should include listing time periods that reflect anticipate progress, the methods and skills employed to implement them and the cash flow associated at each period.

A simple bar chart can be used to complete this task. This would comprise a chart where the bullet point planning activity as given in 9.3.2 are listed on the left hand side of a chart and the months of the year along the top of the chart. Note that separate charts will be required for each market segment.
It must also be borne in mind that the previous sub chapter may be subject to ongoing external statistical data in respect of market data obtained from statistical authorities. Where such circumstances arise it may be necessary to use what can be termed as an iterative approach. Put simply this will require business owners and managers to revisit certain marketing plan factors.
This in turn could result in a redefined objective that will modify a strategy with a consequent impact upon the planning and implementation of the plan.

It must be observed that any implementation programme may consist of multiple departmental activities. This being so, an overall programme should be provided as well as any individual departmental activities. Here the onus of producing the latter would rest with the relevant departmental supervisor.

Clearly, the process considered within this separate chapter may be preceded by a report on any activities associated with previous planned improvements undertaken within the past trading activities.

Finally, we are warned by text book authors that it is illogical to plan and implement strategies that are not rooted within the capabilities of a given business activity. The underlying goal it is suggested is to anticipate any implementation barriers as early as possible.

9.3.4 Control

In its simplest format control is the means employed to measure the output of the *business plan* and its impact upon the business and to take any corrective actions to ensure the objects are being met. However, it can also be said that it is the factor employed to measure the marketing impact upon the segment within which you operate that offers value to the customer while creating an acceptable level of financial return for a given business owner or manager.

To be successful in the control of a marketing plan and a *business plan* there is a need to understand the overall business sphere of activities and not just the day to day running of the business. There is a need to construct a simple method of collecting the data that is required to support and implement the marketing plan. To observe any best practise techniques employed by others, be they competing or supporting organisations. Indeed, all of these requirements can be summarised as constructing an information network that allows a simple method of importing data, storing data and exporting data. It should be noted that any descriptive data provided within the *business plan* that relate to the management techniques employed within the company be it current or a projected need must be undertaken in a manner that will not disrupt working relationships.

9.4 Chapter Summary

The overall objective of this particular chapter will be to ensure that the projected P&L is not merely achievable but has the potential to improve in margin terms. It follows that various methods of measurement will need to be assimilated within the business. There is a need to measure the progress associated with achieving the corporate objective. How are our strategies functioning to support objective needs? How do the needs of this chapter support the strategy review process?

An analogy would be R&D effort to produce a new microprocessor for NPD. The engineer would be engulfed with algorithm analysis. His or her manager would want to know if it is on programme. Whereas the business owner will want to know if it will work, are the costs on schedule and can we have it sooner at a lower cost?
Conversely, the marketing plan would ask all of these questions but would also consider the same impact upon the marketing services paradigm stated earlier.

Chapter Example

Having completed our financial analysis of the *business plan* we must now review the strategy we wish to pursue to successfully implement it within our business micro environment. We must also observe macro environmental factors that may have a direct impact upon any strategy we would wish to employ.

It is the needs of this particular chapter that has given rise to the observation provided within chapter 10.0. In that we must concern ourselves with implementing the *business plan* as well as the strategic needs associated with implementing our marketing plan. Here again we have responded to the iterative process employed throughout.

For the purpose of this example we will put forward a simple statement at the outset of this sub chapter that alerts *business plan* scrutinisers to our intent; fig 42 refers.

Fig 42
Strategic Analysis – How are we going to get there?

The Marketing Plan Planning Process"

"Business Plan Implementation Programme"

"Strategic Intent"

Page 1 of 2

The strategic intent suggested by fig 42.1 is derived from a mix of that suggested by Table 9A given above.

Fig 42.1
Strategic Analysis – How are we going to get there?

The Marketing Plan Planning Process"

"Strategic Intent"

To identify the association between business plan operational and long term strategy implementation in respect of marketing plan implementation.

Page 2 of 2

The latter sheet could be expanded to include operational and long term maximisation of opportunities and the impact it will have on limiting the management of the risks respectively. This alone is somewhat of a broad statement that will require further definition. One cannot state that something will be implemented without any form of acknowledgement for its need. The following two examples (fig 43 & 43.1) are used to describe our proposals for IVS.

Fig 43
Strategic Analysis – How are we going to get there?

The Marketing Plan Planning Process"

"Strategic Intent Process Employed – Micro Environment"

Market Intent

- A realignment of manufacturing and overhead core competences to sustain operational market leader status but to enhance this status for long term corporate strategy

Business Intent

- To implement a management structure that will separate performance measurement from sales process management

Page 1 of 2

Fig 43.1
Strategic Analysis – How are we going to get there?

The Marketing Plan Planning Process"

"Strategic Intent Process Employed – Macro Environment"

Market Intent

- To increase market entry barrier levels by incorporating our own specific marketing plan

Business Intent

- To adapt and streamline our sales processes further to limit all of the risks arising out of the increased opportunities

Page 2 of 2

Here we are able to determine that our implementation programme will not only impact upon the business but the macro environment within which it operates. Conversely we can also suggest that the statements made will allow the company to balance its "technology push" requirements with the macro environmental "market pull".

Support for this suggestion is observed as an area of focus when using relationship marketing techniques. To this end it is an area of observation where there is a need to focus upon customer needs throughout the segment. Dixon & Wilkinson (1984, P.62) also review segment analysis in a similar vane but add that;

> "the basic idea underlying market segmentation is that any market is likely to consist of sub markets which might need separate marketing mixes"
>
> Source: McArthy, E. J (1981, P.224). Basic Marketing: A Managerial Approach, Homewood, Illinois, Richard D. Irwin.

This then implies that whilst there is a need to consider the micro environmental needs of a given marketing strategy one must also consider the needs of the overall segment. It is to be noted that this could in fact be assessed in a reverse manner where the needs of the overall segment must consider the micro environmental needs. While this may border on "technology push" and "market pull" definitions it is not the intention to review this fact any further at this stage of our example.

Thus we could now suggest that the ontological needs of this sub chapter are complete but that we must now consider our process of implementation. Before we begin this process we must be sure of any ambiguities within our *business plan* that may have a direct or indirect impact upon the process employed to arrive at our strategic intent. This we will define as our epistemological evaluation. The ambiguities referred to are those that may influence opportunities and risks.
Let us not forget that the *business plan* documentation put forward including any appendices will be used as a point of reference by those who has the responsibility of managing its implementation and by those who will manage its performance measurement.
Our goal for any retrospective activity at this time would be to reanalyse factual data as well as any CUGs employed to confirm or otherwise that which we wish to implement represents the true potential strength of the company in respect of known markets or those we may create.

This example would suggest that we have three areas to focus upon (*Analysis Phase*) that impact upon our implementation programme. These are:-

- To convert IVS from a "sales" based organisation to a "marketing" based organisation.
- To analyse the impact of the marketing plan in relation to inter department existing sales process activity.
- To analyse the impact of strategic intent (chapter 7.0) in relation to inter department existing sales process activity in terms of P&L objective requirements (chapter 8.0)

Individual company needs may suggest that different analysis investigation techniques must be used. However let us not forget that one of our main aims here is to inform or agree with others what is required of them to fulfil the needs of the *business plan*. There will be those who have no commercial bias who will potentially provide barriers to the implementation programme.

Most probably we will need to present or discuss our needs in manner that will make it more appealing to these individuals. Mahatoo (1989) suggests that there is a need to create a stimulus that responds to the psychological state of the individuals under review.

The use of operational and long term objective criteria for this example together with the manner in which it is presented will have a direct association with our implementation programme. Any change to that which exists already one would argue must not be radical in nature to the extent it will disrupt operational sales objectives.
Rather they should be examined such that any change can be implemented gradually with little or no turbulent effect. As with other sub chapters there are no set processes we can examine to apply to individual company implementation programmes.

This being said it could be argued that management experience would suggest that useful guidelines can be provided; fig 44 & 44.1 refers.

Fig 44
Strategic Analysis – How are we going to get there?

The Marketing Plan Planning Process"

"Implementation Programme"

Page 1 of 2

Fig 44.1
Strategic Analysis – How are we going to get there?

The Marketing Plan Planning Process"

"Implementation Analysis Phase"

- To convert IVS from a "sales" based organisation to a "marketing" based organisation
- The marketing plan in relation to inter department existing sales process activity
- The impact of strategic intent in relation to inter department existing sales process activity in terms of P&L objective requirements

Page 2 of 2

The first bullet point has the potential at the outset of this sub chapter to create confusion for those not familiar with the difference in the nomenclature given. Consequently we may be faced with a difficulty in planning of our "implementation process" requirements. It is suggested therefore that we use this potential for confusion to simplify the process of implementation.

If we present the *business plan* as our primary business objective it will give rise to an increase in the knowledge base in respect of how the output of one department will impact upon another and its corresponding impact upon business performance. While this inter departmental interactivity is more associated with a sales process it is the marketing activity that has created a need for a sales process to be implemented.

Put simply it is suggested that it is the marketing plan that creates a point of origin for *business plan* implementation planning. Therefore, for our example we will provide four separate sheets that identify which strategic planning techniques will be employed by IVS

Fig 45
Strategic Analysis – How are we going to get there?

The Marketing Plan Planning Process"

"Business Plan Planning – Focus of Evaluation"

Page 1 of 4

To convert IVS from a "sales" based organisation to a "marketing" based organisation	
Focus	Objective
Product/Service	Operational ▪ Marketing plan selection and implementation. Long term ▪ Review marketing plan changes in respect of macro environmental factors.
Aims/Strategy Analysis	Operational ▪ Measure impact of marketing plan implementation by way of market share/opportunities and financial results and efficiency factors Long term ▪ Revisit R&D strategy and modify accordingly in respect of marketing plan implementation and changes arising out of macro environmental factors.
	Page 2 of 4

Fig 45.1

The impact of the marketing plan in relation to inter department existing sales process activity	
Focus	Objective
Product/Service	Operational ▪ Analyse marketing plan impact upon existing sales process & process modifications. Long term ▪ Measure impact of the marketing plan in respecting of company Position/Brand
Aims/Strategy Analysis	Operational ▪ Identify resource expenditure cash flow requirements. Long term ▪ Identify suitable process for measuring impact of implementing critical success factors.
	Page 3 of 4

Fig 45.2

The impact of strategic intent in relation to inter department existing sales process activity in terms of P&L objective requirements	
Focus	Objective
Aims/Strategy Analysis	Operational ▪ Review existing performance measurement techniques in relation to marketing plan objectives analysis Long term ▪ Identify a programme for critical success factor implementation that does not impact upon sales output. ▪ Identify programme of implementing the output of critical success factors analysis & implementation across the market segments
	Page 4of 4

Fig 45.3

It will be noted that the planning aspect of our *business plan* implementation programme as given within these four sheets does not delve into any specific detail. The purpose of this is to remind the management team of what it is that it must plan for in respect of *business plan* effectiveness. They in turn would need to discuss with their teams the finer details of the techniques to be employed to achieve an end to an aim.

Here the latter aims are those we identified within chapter 7.0. They are those we must implement to the product, place, and price and promotions policy.

However, as can be seen the *business plan* planning aspect of our *business plan* implementation warrants a review of the 4Ps not only in respect of the marketing plan but also the sales processes employed by IVS. It follows that the implementation techniques to be discussed by departmental managers and their respective teams undertake an analysis of how their actions will impact upon the sales process as given within the following example (fig 46 & 46.1)

Clearly, the number of objectives set would suggest we exclude all of them for this example and that we provide an example comment. In bar chart format a typical presentation could consist of:-

Fig 46
Strategic Analysis – How are we going to get there?

The Marketing Plan Planning Process"

"Business Plan Planning – Programme of Implementation"

Page 1 of 3

Product Policy Implementation – Fig 46.1					
Aim	Sales Process Impact	Activity Analysis	Start Date	End Date	Responsibility
Implement new control system technology	Operational	Undertake design & test for selected markets	May	January	Jonathan/Manfred
	Long term	Create new manufacturing materials list and amend manuals	September	April	

Under normal circumstances we would complete the product policy implementation with all the remaining strategies that were identified. Similarly, a bar chart would be provided for place, price and promotions policy strategic aims.

Furthermore we must plan for implementing a marketing orientated management focus. Our analysis as given above would suggest an additional bar chart that could be constructed as follows:-

Conversion of Sales to Marketing Operations Philosophy Fig 46.2					
Aim	Sales Process Impact	Activity Analysis	Start Date	End Date	Responsibility
Implement new business plan	Operational	Off site meeting to agree content	March	April	Eugene
	Long term	Barrier identification and procedures to be employed	March	July	Eugene

Owing to the number of policy objective criteria this sub chapter has the potential to be somewhat detailed in format. This would suggest perhaps a cursory glance by others. Should this be so we must bear in mind that irrespective of volume of content all data will be retained as part of the business operations philosophy. Accuracy of information is crucial as it will have a direct impact upon the methods employed to control our planning techniques.

There are many software programmes available that contain a mix of algorithms that drive myriads of complex equations to ease a given specific business process. They are packaged in a similarly numbered mix of solutions. Their usage will have been sourced using a process where one particular company is seeking a certain solution or where a vendor has sourced the product through unsolicited means.
If we analyse the process we have undertaken to arrive at this juncture of our *business plan* then how can we be sure that the computer based equipment supplier has undertaken the same process. The net result it could be argued is that somewhere within the usage of the product IVS would need to adapt its own processes to align with the features employed to operate the computer based process. This situation has the potential to introduce deficiencies within a host user.

If we use a modern car as an analogy we can state that the advent of technology has allowed a more efficient method of preventative maintenance and eventual breakdown. Even though engine technology has advanced significantly there are still numerous sensors that are required to monitor engine performance. It could be argued that while we do have an improved engine management system that our most common failure would be associated with a sensor that is not functioning correctly.
The reason for using this argument relates to any measurement systems we must employ to implement the control of our *business plan*. Complexity could be attractive if it improves efficiency but even then

efficiency for one department may not have a positive impact upon the sales process. This would suggest that there must be basic needs that must be fulfilled. Any supplier benefits over and above these basic needs can then be assessed by way of their impact upon the *business plan* if any.
So what are our basic needs? Here the business plan would suggest three factors for consideration:-

- Measurement of overall *business plan* objective criteria
- Measurement of marketing plan performance criteria
- Measurement of *business plan* impact upon existing sales process

The latter bullet point given above is particularly important in that *business plan* objectives must be put in force in a manner where it can be reproduced repetitively irrespective of the person employed to manage the process adopted to produce it. Thus we could argue that the first bullet point would consist of a system that would measure strategic directional planning policy from a departmental activity point of view and its impact upon the profit and loss statement and cash flow objectives. The intended recipients of data would be the management team.

The second bullet point would consider a process where any change in market audit data would be stored and its impact upon SWOT analysis and strategy formulation identified for any review in *business plan* operating philosophy. Here we must not forget that the *business plan* is produced using data at a given point in time. Changes in market environmental factors will occur; albeit any such changes may not necessitate a change in company operational strategy. Long term strategy could be affected hence its importance.
The intended recipients of data would be departmental managers/supervisors.

We can summarise these three criterions for this example by way of a simple table for presentation purposes. It is proposed that a two tier system be introduced that will allow a separate analysis of corporate strategic needs and operational sales process activity, Table 9.1 refers.

Control Measurement Criteria	Operational Strategies	Long Term Strategies
Analysis of Business Plan Performance	Executive analysis measurement and control	Executive analysis measurement and control
Analysis of Marketing Plan Composite Data	Management analysis and change control	Executive analysis measurement and control
Analysis of Sales Process Performance	Management analysis and change control	Management analysis and change control

Table 9.1

Supporting business plan research criteria

Supporting text for the example provided does exist but not in the form of operational and long term objective criteria. McDonald (2003. p472) states:

> "Performance against plan for product, market, channel and so on can be tracked through simple graphical display, or through facilities such as exception reporting, where the system highlights areas where the divergence from plan is significant".

Clearly, any divergence suggested by McDonald would by way of that put forward within this *business plan* be identified in terms of their impact upon operational and long term strategic needs (marketing plan), marketing plan repetitiveness and adaptability (sales process) and *business plan* objective criteria. Kotler Et Al (1997. p. 105) would seem to group the primary aims of this chapter with an overlap of that we will review within chapter 10.0.
Jobber & Lancaster (2004. p.45) discuss a similar method of control although without the use of a specific model but with additional guidance in the form of linking the measurement process as a feedback loop to the "Formulate Strategies" marketing plan criterion.
Such a feedback loop will ensure that it is the strategies employed and these strategies only that the company must concentrate on.
None of these authors have acknowledged the need to create an additional box between "Set Goals" and "Measure Performance" as indicated in figure 47 below*. To this end, there would need to be a suitable management structure in force that not only understands the criteria associated with the goals set but also are able to effectively evaluate corrective action and corrective action. Chapter 10.0 considers this potential void within existing text book consideration.

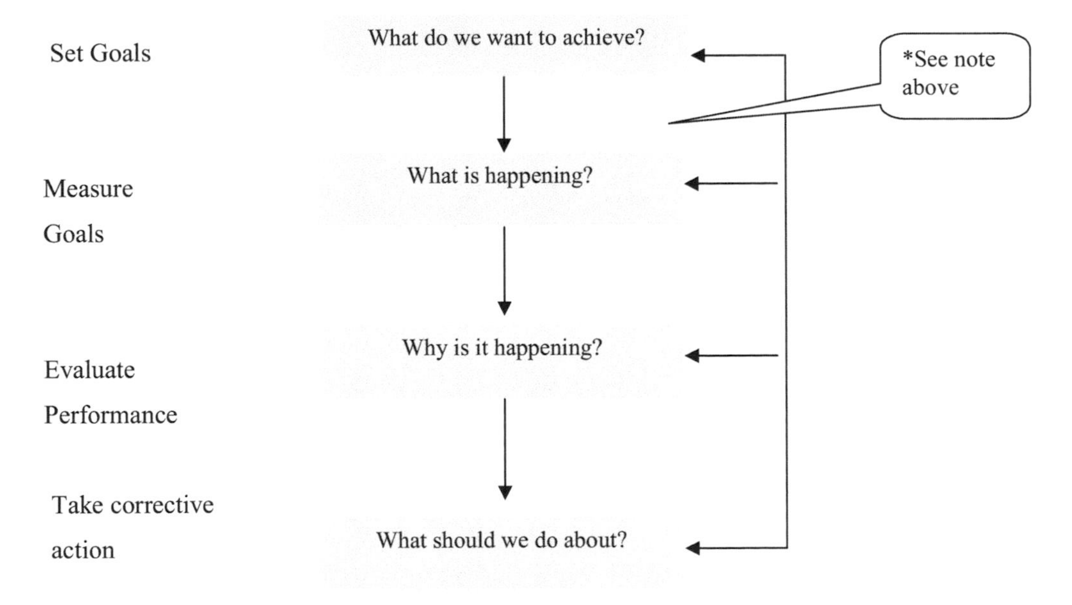

Fig. 47

Business Performance Review & Evaluation

All businesses must have some method of reporting good or poor performance. A ***business plan*** should consist of a description of what methods will be employed to undertake such a requirement. The methods used must be able to reflect upon previous trading performance as well as those methods used or anticipated for operational activities. This we can split into two factors:

- Financial performance
- Marketing performance

Similarly there exists an argument that the performance of any company and its good or otherwise impact is monitored by the market (Kotler et al (2002)). It follows that any performance review must be undertaken bearing in mind its impact on the environment.

The objective for this chapter is to simplify governance of the business but in an efficient manner. This is gauged by way of the methods of reporting employed, by whom and in what way. It will be seen that market performance will impact upon financial performance. Similarly, the latter through business performance (*internal factors*) could have an affect on market performance.

To all intents and purposes it is the marketing implementation programme as discussed within the previous section that will determine what it is that will require some form of performance measurement. However, to maximise the level of efficiency and accuracy of the performance data provided, it was suggested that internal marketing be employed. Given that most businesses will have not yet contemplated such an activity we will review some of the factors that will impact upon governance as adapted from Piercy N.F (2003).

- Identify influential supporters and opposition for the corporate objectives employed.
- Ensure that the influential supporters form part of the business decision process.
- Develop implementation strategies for the marketing plan (or its derivative).
- Identify where any barriers to the latter will occur.
- Who are the process owners of the implementation programme and what is their level of empowerment?
- Who is it that focuses on exploiting the capabilities of the existing organisation in a way that it will match company strategy? What barriers to implementation are known to exist?

There will then be a need for the influential support management team to effectively manage their activities and to report accordingly but in a manner that interacts on a cohesive interdepartmental basis.

Put simply the author would argue that governance can be represented in the form of a basic paradigm as shown in Fig 48

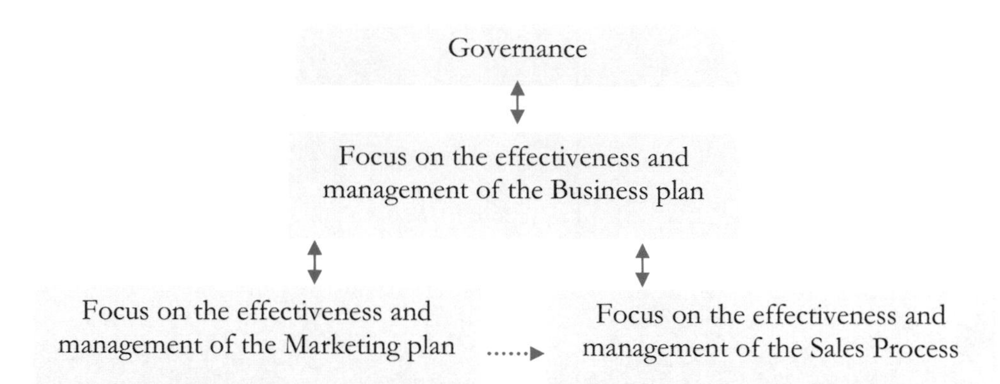

Fig 48 Suggested Company Governance Management Structure

10.1 *Business Plan* Management

Here the focus of management and reporting activities is suggested in the form of an *offensive business analysis* Donaldson, H (1997). Depicted as Fig 49 the latter is a particularly suitable paradigm for use of reporting on performance in several ways.

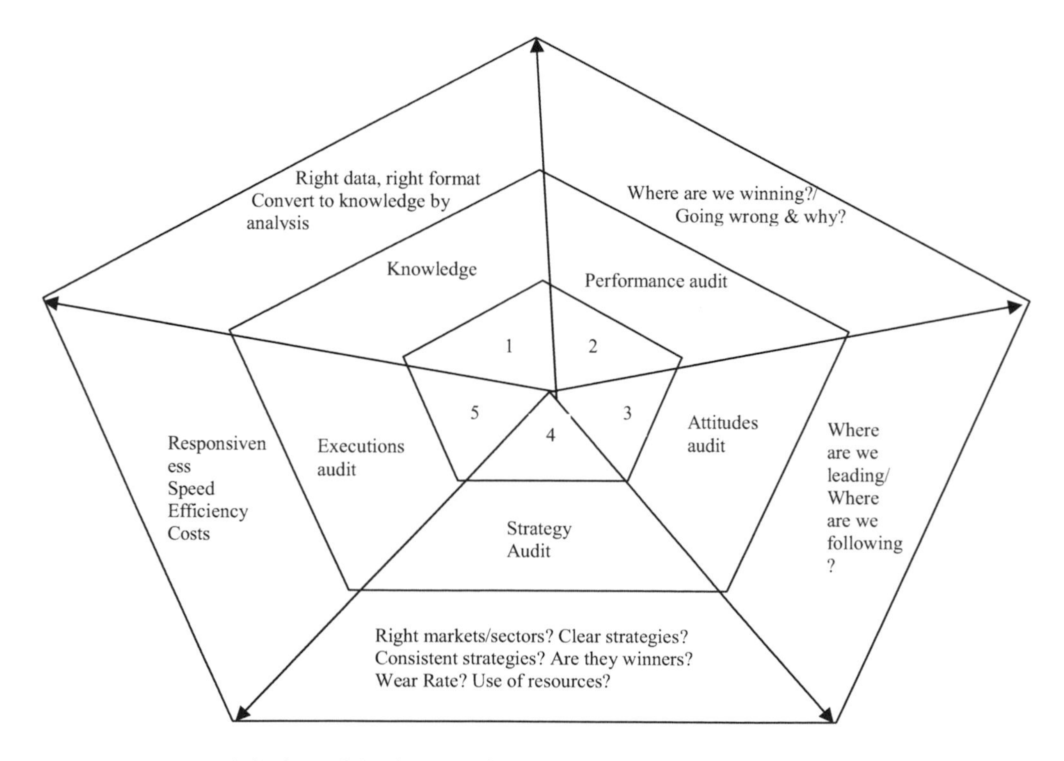

Fig 49 Modified model of internal examination (Davidson, H (1997)

Its major benefit is that any monthly or otherwise reporting structure is undertaken using macro environmental factors and not simply "This is what we did last month"
The word "audit" is seen to be used but this can be replaced with appraisal, assessment, examination or simply "report" or whatever is preferred when producing the overall analysis.

10.1.1 Knowledge

There is a need to convert facts and knowledge into understanding through analysis. Useful tools for effective analysis are given as right data, systematic approach, focused approach, be thorough and probing, must produce an action not a delaying tactic and must lead to continuous improvement.
He goes on to suggest that this first stage is extremely important as it will act as the basis for making judgements in the following four stages.

10.1.2 Performance audit

10.1.2A Market Performance

There will be a need to provide statements in respect of market success, market failure and why? Sub chapters that will support these statements are the facts associated with segment activities, sales channels, consumers (users), customers, product/s, innovation or NPD (new product development) and profits for instance. That is in what way do these factors impact upon each other?
To all intents and purposes any questions and answers generated within this audit will assist with the identification of future opportunities and areas for improvement – will be needed for stage 4.

10.1.2B Financial Performance

Certain companies will require or use as part of the *Business Plan* a number of statements or terms that relate to business performance. The terms given within this sub chapter (Table 10.1) are put forward for use in this regard as adapted from Baker (2000) and CUG provided by others (D. Noble). Needless to say, some of them are of an accounting nature such that the responsibility for their production should rest with the finance department. Here, the objective for review is the performance criteria put forward in relation to previous trading activities and how the *business plan* will be shown to improve or sustain the ratios provided.

Financial Business Glossary	Explanation
An Asset	Anything or everything owned by the company for use for providing a benefit at any given time. They can be split into:- • Fixed Assets + Tangible: Land, Buildings, Machinery + Intangible: Goodwill, Brand Names + Investments: Possibly in external institutions

	• Current Assets ◆ Stock: The value of material at any given time that will be used to produce a sale ◆ Debtors: Monies owed to the company as well as the value of down payments to others for goods plus the premiums for any valid cost of sale.
A Liability	Anything or everything owed by the company to an outside source of funds or goods. They are the liabilities associated with the trading activities for the given year. They can be split into:- Current Liabilities ◆ Creditors: Amounts owed as the cost of trading such as materials and employees. ◆ Loans: Amounts due to any financial institution arising out of a loan or otherwise required to sustain the *business plan.* ◆ Overdrafts: This is the balance outstanding at a given time. Long Term Liabilities ◆ Long term Loans; Those loans needed that are repayable over one year. ◆ Leases: Cost of any leases greater than one year. ◆ Deferred Costs: Tax or other liabilities the company may be required to pay. Shareholder Funds Those companies who operate using shareholder funds will be required to operate using share capital and share reserves.
Balance Sheet	This is a simple statement used to show the financial status of the company by way of identifying the value of all of the assets and liabilities at any given time.
Performance Ratios	Normally used to provide an indicative status of management performance. Several ratios are used, typically these will consist of:- Percentage Profit on Sales

$$\text{Margin} = \frac{\text{EBIT}}{\text{Sales}} \times 100$$

Here EBIT is the acronym for Earnings before Interest and Tax. The word earnings is often substituted with Profit thus the acronym would be PBIT. Either way the formula will provide a simple analysis of the margin generated by a sale.

Percentage Return on Capital Employed (ROCE)

$$\text{ROCE} = \frac{\text{EBIT}}{\text{Capital Employed}} \times 100$$

Capital employed is the total of all long term liabilities and any provisions at any given time. It can also be the sum of all assets less the current liabilities that are used to sustain business activity. Put simply the ratio provided is used to evaluate the profit generated in respect of the value of the net assets.

Asset Turnover

The ratio provided is used to analyse the level of sales the company is able to produce from a given level of capital employed. It can be compared to the interest the company would generate by having the same capital lodged within a financial institution.

$$\text{Asset Turnover} = \frac{\text{EBIT}}{\text{Capital Employed}}$$

Text book data tell us that there are options for comparing the relationship between ROCE, margin and turnover. This is expressed in the following manner:-

$$\text{ROCE} = \frac{\text{EBIT}}{\text{Capital Employed}} \times 100$$

$$\text{ROCE} = \frac{\text{EBIT}}{\text{Sales}} \times \frac{\text{Sales}}{\text{Capital Employed}}$$

ROCE = Margin x Asset Turnover

The author would suggest that latter equation is useful as a business performance indicator if fully understood by the responsible process managers. However, it does not reflect the pace at which the volume of assets is traded and thus must be recognised when producing the ratio.

Stock Turnover

Capital employed is very much an issue when reviewing company stocks, i.e. the materials or otherwise held in order to produce a sale. The text book equation used to provide an evaluation in respect of this problem is:-

$$\text{Stock Turnover} = \frac{\text{Annual Sales}}{\text{Average Stocks}}$$

The ratio is used to measure on an average basis the level of stock holding sales throughout the year. Low values suggest that unnecessary costs are being attributed to the business by the process used to buy and sell the stock. Similarly, if the value is too high it would suggest that the company has insufficient stock required to produce a sale.

This may be stated in a different format (Noble):-

$$\text{Stock Turnover} = \frac{\text{Annual Cost of Sales}}{\text{Annual Stock}}$$

Debtor ratio

Debtors are those that owe the company money. Given that payment is required to sustain business activity prudence suggests the use of a ratio where the collection period used to secured payment is observed. This is normally expressed in "number of days" and is given as:-

$$\text{Average collection period} = \frac{\text{Trade Debtors}}{\text{Sales}} \times 365$$

<table>
<tr><td></td><td>This ratio must be checked in respect of the standard terms of payment forming part of the company's terms of trading. Similarly, they will have further relevance if compared to level of expense incurred by the company to produce a sale.

Thus we have a:-
Creditors Ratio

$$\text{Average payment period} = \frac{\text{Trade Creditors}}{\text{Cost of Sales}} \times 365$$

Further ratios can be presented in the form of measuring performance where all aspects of the business are measured in respect of and their impact upon EBIT divided by the capital employed. These ratios are used to increase ROCE by increasing profit and reducing the capital employed.
Increasing ROCE will require increasing profit by a possible increase in sales and by reducing the costs. Whereas a reduction in capital employed may require a reduction in fixed assets and a reduction in working capital.
In the event that these ratios or similar are required but not used by the finance department that further research be undertaken or contact made with the author who will provide samples for consideration.</td></tr>
<tr><td>Financial Status Ratios</td><td>Here we are interested in the liquidity and solvency of the company.

Liquidity Ratio
Most companies will need to assess that the short-term assets are sufficient to meet short-term liabilities. This is represented in the following manner:-

$$\text{Current ratio} = \frac{\text{Current Assets}}{\text{Current Liabilities}}$$

Here the ratio would imply that the ratio provided will be a value</td></tr>
</table>

greater than 1.0. However, as with some of the ratios provided hereof it is to be noted that the ratio can vary in relation to different industry classification. Any comparison therefore should be undertaken using corresponding values for other companies within the same industrial sector.

Liquidity can also be measured using the "Acid Test" ratio which is similar to the current ratio but where the cost of the stock is excluded from the funds available. It is a ratio which is particular suitable for the manufacturing industry or any other industry where the stock can take several months or more before its use as a sale. Consequently, values less than 1.0 are considered normal however, the value of stock not calculated must form part of the ratio submittal process.

$$\text{Acid Test} = \frac{\text{Current (liquid) Assets (Less Stock)}}{\text{Current Liabilities}}$$

Solvency Ratios

Solvency ratios are often used by lending institutions to assess the level of risk a business would impose if wishing to seek credit terms or otherwise with them. These lending institutions will use a debt to equity ratio which is referred to as "gearing"

Companies often require a high level of funding in order undertake their business activity. In textbook format these are classified as "highly geared" companies. Those with low debt business activities are similarly identified as "low geared"

Gearing ratios are expressed as a percentage:-

$$\text{Gearing} = \frac{\text{Total borrowing}}{\text{Funds available + Total borrowing}} \times 100$$

Both the lending institution as well as the borrower will need to assess the ability of the business to repay any debt based upon company performance. While the actual ratio used may vary it can be expressed as a simple formula, namely:-

	Interest cover = $\frac{\text{EBIT}}{\text{Interest}}$ Last but not least is the measure of profitability. Several arguments exist in text book analogy in terms of how profitability should be measured. Suffice it to say this particularly subject will always remain objective. Useful reading in this regard is John Sizer: *An insight into management accounting*. 1979. (Source: Baker 2000, P.494). Useful extracts are:- Return on Investment (ROI) = $\frac{\text{Net Profit}}{\text{Capital Employed}}$ This is given as the profit attributable to minority and preference shareholders divided by the Equity share holder's investment. However, he goes on to state that other sources have become more specific with their definition by way of:- $\frac{\text{Net Profit (after tax)}}{\text{Total net assets}}$ and $\frac{\text{Net Profit}}{\text{Fixed assets}}$ Gross Profit or gross margin:- $\frac{\text{Gross Profit}}{\text{Sales}}$ and $\frac{\text{Net Profit}}{\text{Sales}}$

Table 10.1

10.1.3 Attitudes audit

Are you a market leader or a market follower? This question must be borne in mind as it applies to the strategy employed and thus, are you achieving your *business plan* aims?
In what way is the implementation programme progressing? What are the internal views on progress (and its outcome) in respect of company employees and market response?

10.1.4 Strategy audit

Using the three previous audit activities one can begin to assess the effectiveness of the *business plan* strategy employed, namely:-

- Is the strategy working?
- What part of the strategy is seen to be effective?
- What part of the strategy is seen to be ineffective?
- What part of the strategy is gaining momentum?
- What part of the strategy is beginning to fade?
- Is the strategy consistent or are there constant (perhaps needless) changes?
- Is there a need to revisit the business plan strategy employed?

This author would suggest that strategy review be undertaken in the form of how the market views your performance, your strategic impact upon the market as well as providing an overview of how the market is performing or functioning.
It is to be noted that market performance can refer to how are they changing? Which markets are growing (by volume or product needs) and which are those that may be in decline? Similarly, in what way are the environmental factors impacting upon market changes (knowledge base)?

Market views on business performance is an under valued maxim by many industries. Those companies that do not employ a marketing culture are unlikely to respond to these needs in the same way as marketing based organisations if at all. The overall impact of the latter is far reaching as can be seen within Fig 50.

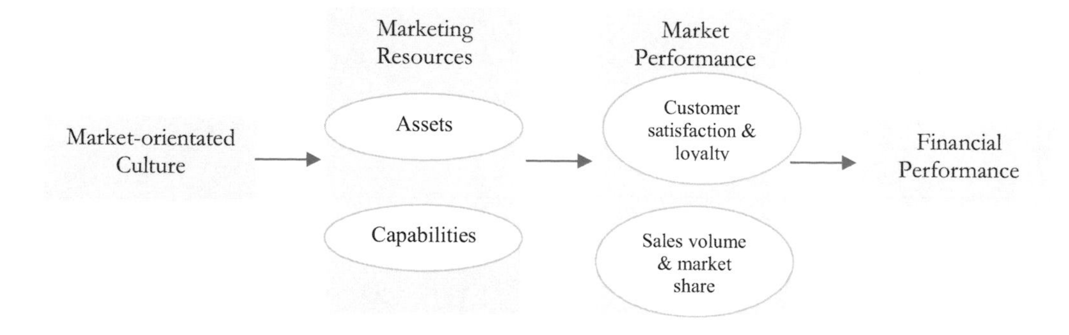

Fig 50 Marketing and performance outcomes (Hooley, et al 2004)

10.1.5 Execution audit

It is stated that companies should aim to win on the quality of the strategy as well as the quality of its execution. He suggests that excellent execution can transform an average strategy into a winning formula. Key factors for responsiveness consideration are:-

- Customer needs Vs competitor initiatives
- Speed and efficiency of implementation
- Cost position

This *business plan* would suggest that "quality of execution" could be further expressed by way of the need to examine existing sales processes and any impact arising out of its implementation. Additional bullet points would serve to describe potential areas where deficiencies could occur.

- Cash flow Vs Value/Volume
- Existing product/Existing market process efficiency analysis
- Existing product/Existing market/Market penetration impact on process
- New products/New markets impact on process

In concluding this particular sub chapter we could argue that a report on activities be provided for each individual aspect of the business such as sales/competitive trends, manufacturing activity, finance activity and customer relationship trends.

10.2 Marketing Plan Management

The purpose of this chapter is to simplify the identification of the tools that are available or required to generate a marketing plan. Ergo, what is it that needs to be measured and what is its impact upon the *Business Plan* planning strategy?
In the first instance a marketing plan overview can be assessed in the form of a summary as depicted earlier in Fig 11, Page 30.

It will have been observed that this plan has ten main constituent parts and that these in turn have many different factors that provide a means to an end. Marketing as a science provides substantially more data on all of these individual factors as well as others for a varying degree of industry types and individuals. Those wishing to undertake further research in respect of these factors should seek text book support or alternatively check what is available via the internet or contact this author. Fig 51 can be used to simplify the process of research requirements.

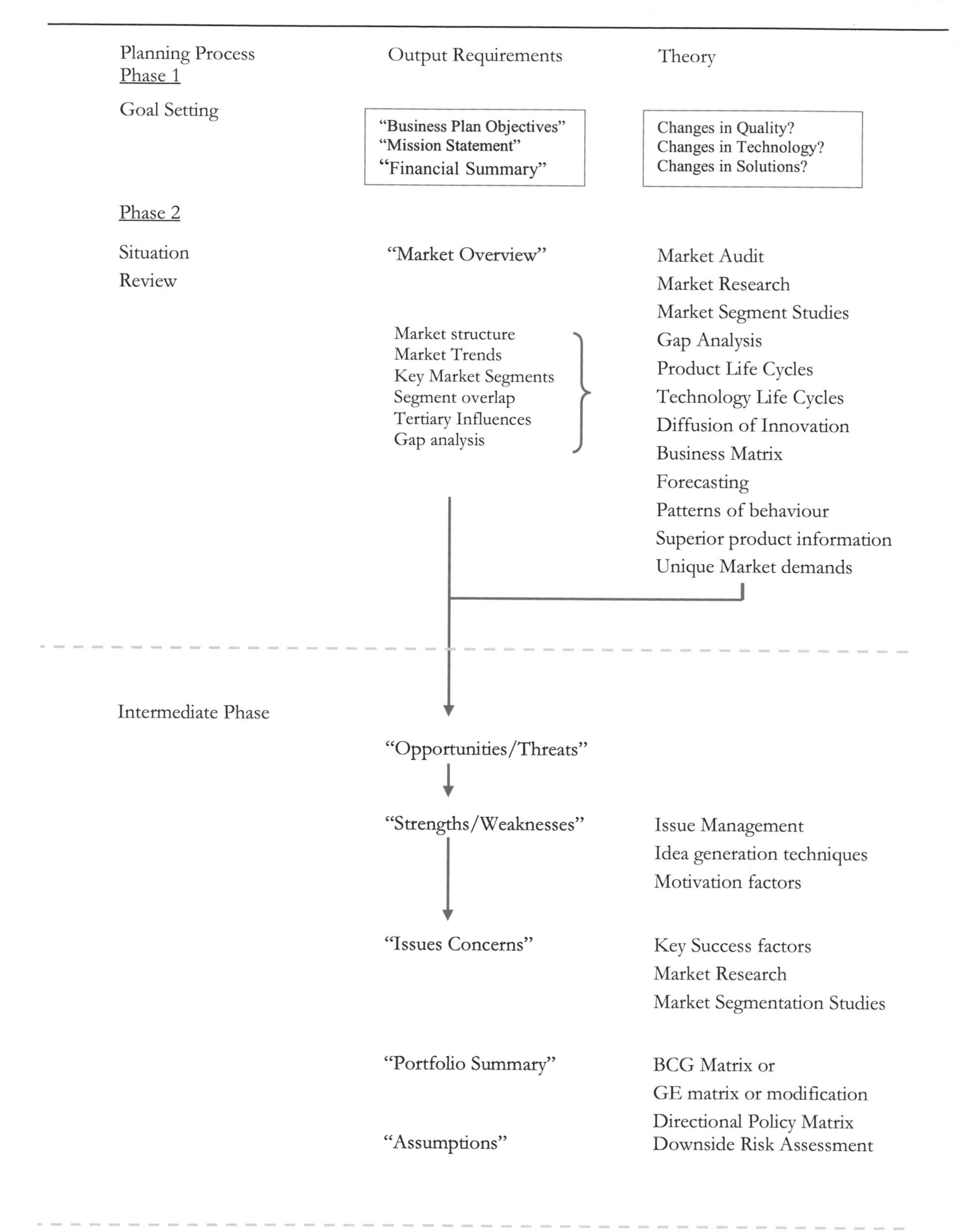
Planning Process
Phase 1
Output Requirements
Theory
Goal Setting
"Business Plan Objectives"
"Mission Statement"
"Financial Summary"
Changes in Quality?
Changes in Technology?
Changes in Solutions?
Phase 2
Situation
Review
"Market Overview"
Market Audit
Market Research
Market Segment Studies
Gap Analysis
Product Life Cycles
Technology Life Cycles
Diffusion of Innovation
Business Matrix
Forecasting
Patterns of behaviour
Superior product information
Unique Market demands
Market structure
Market Trends
Key Market Segments
Segment overlap
Tertiary Influences
Gap analysis
Intermediate Phase
"Opportunities/Threats"
"Strengths/Weaknesses"
Issue Management
Idea generation techniques
Motivation factors
"Issues Concerns"
Key Success factors
Market Research
Market Segmentation Studies
"Portfolio Summary"
BCG Matrix or
GE matrix or modification
Directional Policy Matrix
"Assumptions"
Downside Risk Assessment

Phase 3		
Strategy Formulation and Idea Generation (Balancing criteria)	"Marketing Objectives" (By product segment & Product)	Business matrix BCG Matrix or McKinsey/GE Or modification Directional policy Gap analysis Specific model analysis Product improvement process
	Strategic Focus Product Mix Product Development Product Deletion Market Extension Target Customer Groups	
	"Marketing Strategies" (By 4Ps/Positioning/Branding)	Existing strategy review Market Segmentation Studies Market Studies Competitive Strategies
Phase 4		
Competence Allocation and Monitoring	"Resource Needs"	Forecasting/Budgeting (P&L Statement) **MEASUREMENT & REVIEW**

Fig 51 Strategic Marketing Model adapted from McDonald, M (2003)

Phase 4 of the marketing plan will need to be evaluated using measurement and control metrics that can be used to support the business plan reporting structure discussed within the previous section. Clearly, company governance cannot be efficiently undertaken if reams of documents must first be assessed. It is anticipated that the methods employed will vary depending upon a given industrial sector coupled with personal preference. That being said, certain marketing models can be adapted as suggested throughout this book.

10.3 Chapter summary

This chapter will to a large extent provide definitive guidance for those who set the challenges and answers for many of those often asked "why" and "so what" The same chapter can also act as a top level assessment of the training needs of an organisation. It will allow company owners to separate the training needs for the management team as well as all other "sales" process based employees. To this end, why expend monies on sales training needs when business unit managers do not respond to market feedback?

Chapter example

This example has the potential to impact the most upon IVS. A greater level of empowerment has been identified as part of our corporate strategy and by directional policy evaluation.

In essence this chapter completes the gap identified in model 9.5 depicted within chapter 9.0 as well as allowing a more in depth analysis of the remaining model criterion.
Figure 48 above would seem to reflect more than adequately all that this chapter would suggest. It defines the level of empowerment to be employed by way of responsibility of management review and evaluation. Used in conjunction with figure 49 we can begin to disseminate what it is we wish to review and onus of responsibility. Moreover, it will identify what changes if any that will occur within the IVS business environment that they will need to adopt, approve or otherwise.
It should be noted that we are still at the *How are we going to get there?* phase of our *business plan.* One cannot state *increase level of empowerment* as an objective without first stating what the level of empowerment must focus upon. It is for this reason that our first presentation sheet (fig 52) used for this example uses marketing principles as a guide for our focus of evaluation in respect of the latter.

Fig 52
Strategic Analysis – How are we going to get there?

Business Plan Performance Review & Evaluation

"Governance evaluation"

- To employ techniques that will allow us to monitor, grow and make more efficient our knowledge base
- To identify and implement an efficient Performance analysis system
- To modify IVS business attitudes to include macro environment factors
- To operate with clear and consistent business & marketing strategies
- Sales process re-evaluation in terms of modified operational and long term objectives

Page 1 of 1

Here the criteria associated with these five bullet points are those which are functional to the company. They do not state or suggest the manner in which they must be implemented, the onus of review or the onus of final responsibility. To all intents and purposes this brings us to the final *business plan* question *what will we do when we get there?*

Our analysis of the figurative presentation of figure 49 as well as its subject matter contents would suggest for this example that it can be presented in two different ways as part of the *business plan* presentation techniques to be employed. In its direct application format we could provide a darker shaded area as executive governance (fig 53 overleaf). This implies that closer to the hub one gets they will be concerned more with executive review and evaluation rather than the day to day running of the business. The opposite would apply to departmental/process management activity. For example Hayley would have a responsibility for the performance audit for her department. The performance data would be assessed by one of her team but it would be Manfred and/or Eugene that has the responsibility to agree to any changes (arising out of the audit) suggested by Hayley.

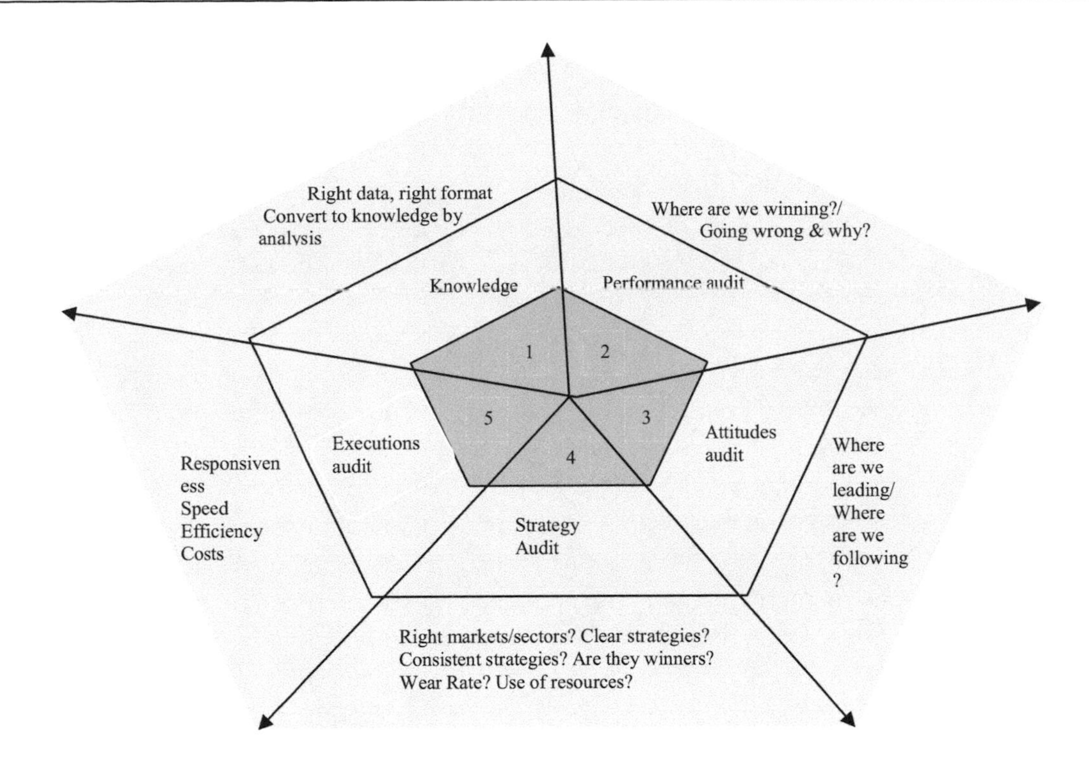

Fig 53 Modified model of internal examination (Davidson, H (1997)

Conversely, the same model could be represented using 5 individual triangles and to then expand accordingly; an example of one such triangle is given as figure 53A.

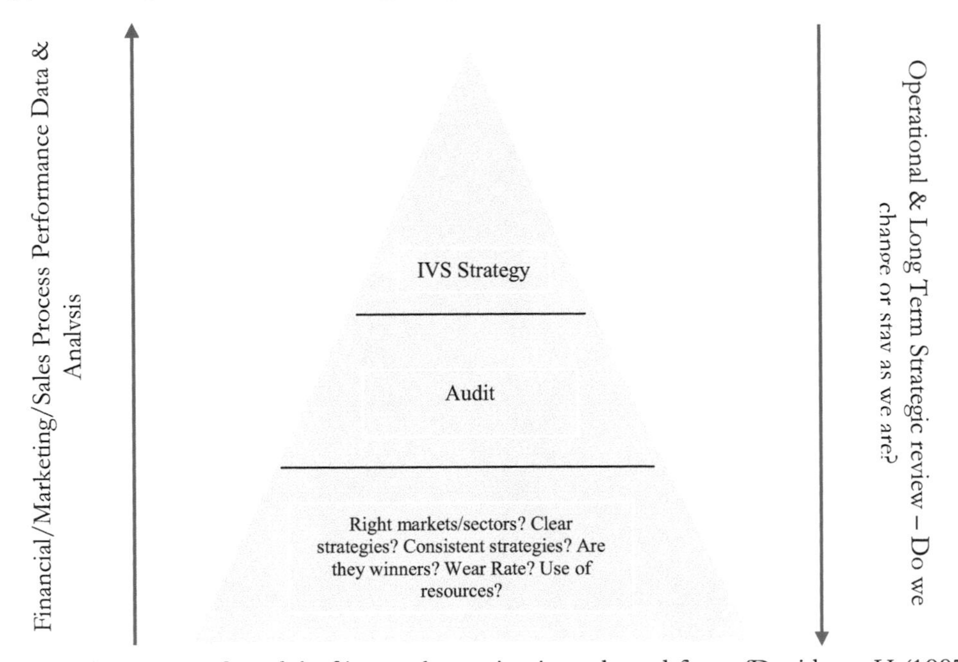

Fig 53A Extract of model of internal examination adapted from (Davidson, H (1997)

Irrespective of the presentations techniques employed they should all be summarised in a way that reflects model usage as being associated with all company departments. For this example, each departmental manager would report every month on the activities they have undertaken in respect of each of the five objective criteria put forward.
However, they would also need to review and evaluate on an ongoing basis the empowerment objectives stated within our first presentation sheet given above. Even if the objectives have been completed successfully, constant examination is needed to review and evaluate the impact the objective has on operational and long term needs.

Having assessed *business plan* governance techniques in synopsis format we will need to provide supporting examples of how it will all come together as a functional management package. Thus we will begin as given within the main chapter above with "knowledge".

Data bases now abound in significant numbers and in an equally significant mix of software packages. Clearly, their use is to simplify the storage of data and the professional manner in which data packages can be adapted and presented.

Statistical data would most certainly accompany many of the packages that are available. For this example we must concern ourselves with the collection of relevant data, the method in which it is formatted and its conversion to our knowledge base (Davidson 1997). However, as with our previous chapter it can be seen that any review and evaluation techniques employed must consider a two tier system, these being:-

- *Business plan* (market feedback in relation to our strategies) impact upon operational sales process activities
- Marketing impact upon long term marketing objectives (and its inherent impact in terms of long term sales process adaptation needs)

By having a two tier system it is suggested that any impact upon the two bullet points given above by way of our knowledge assessment can be measured in terms of the *strategy, goals, policies* and *programmes* descriptions put forward within chapter 4.0 (Baker 2000).
To this end, we must be sure that any analysis of the knowledge base must be observed in terms of their impact upon these four latter factors. This creates a focus for providing realistic information for operational business requirements and possibly would put forward an argument that statistics are more suited for use with long term projections.

Consequently we are able to confirm that this particular sub chapter as a first stage of our *business plan* management process is extremely important as a basis for making our judgements for the remaining four stages. Thus supporting the comments put forward by Davidson using the presentation format given in Fig 54.

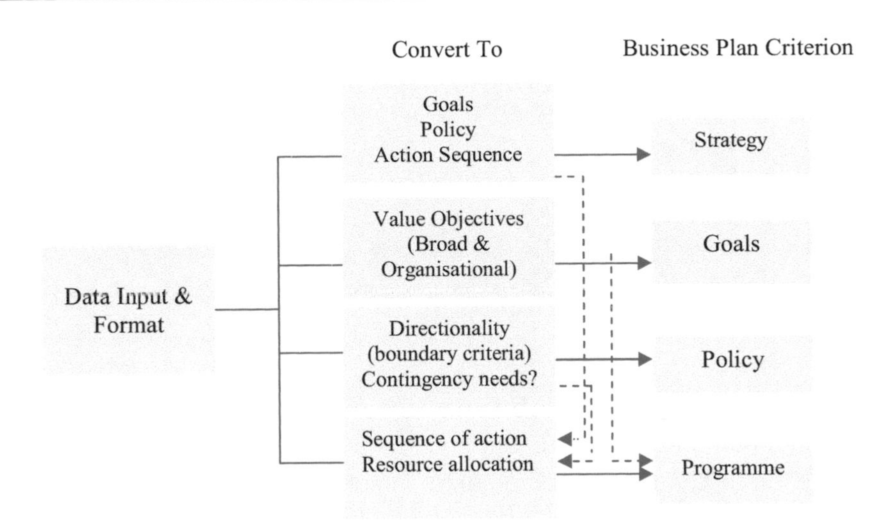

Fig 54 Model of knowledge assessment adapted from Baker (2000) & Donaldson (1997)

A review and evaluation of market and financial performance one could argue attracts the most software packages alluded to. In particular market performance normally sold as a sales management tool. Whereas in fact they are most probably data bases that are used to simplify management audit and consequent action. Rarely will these same packages reflect our 4Ps policy, its channel delivery and personal association with channel activity. Furthermore, there is a danger that such a package could lead to the creation of a separate sales community within an organisation.
To this end, sales individuals are obliged to respond to software package data association with a consequent neglect of corporate strategic intent and sales process objectivity requirements. Ergo, there exist an argument to amend text book "sales force evaluation" techniques depicted.

The example provided as part of this sub chapter would suggest that sale force evaluation should be reviewed and evaluated each month in respect of the objectives set. It follows that its consequent impact upon sales process follower or parallel departmental activity be amended on the same basis by the respective process/departmental supervisor.

The importance of figure 55 (overleaf) is its impact upon the performance revue and evaluation techniques employed. In its adapted format it will be seen to create a point of origin for our knowledge data input/format. This being said many companies will have already expended monies on a given software package and as such our example will consider how we can expand upon a given software product usage in a manner where it is seen to impact upon the financial welfare of the business.

Where are winning or going wrong and why as a text book statement can be expanded from its generic marketing focus of evaluation? The process employed to derive this data and its eventual review and evaluation does not consider its impact upon operational cash flow for a given month or indeed overhead recovery. Similarly, what are our projections for the remainder of the year? It follows that any data must be converted to its impact upon marketing plan, *business plan* and the operational needs of the business activity. The two tier reporting structure stated above will serve to separate knowledge analysis and its corresponding conversion needs.

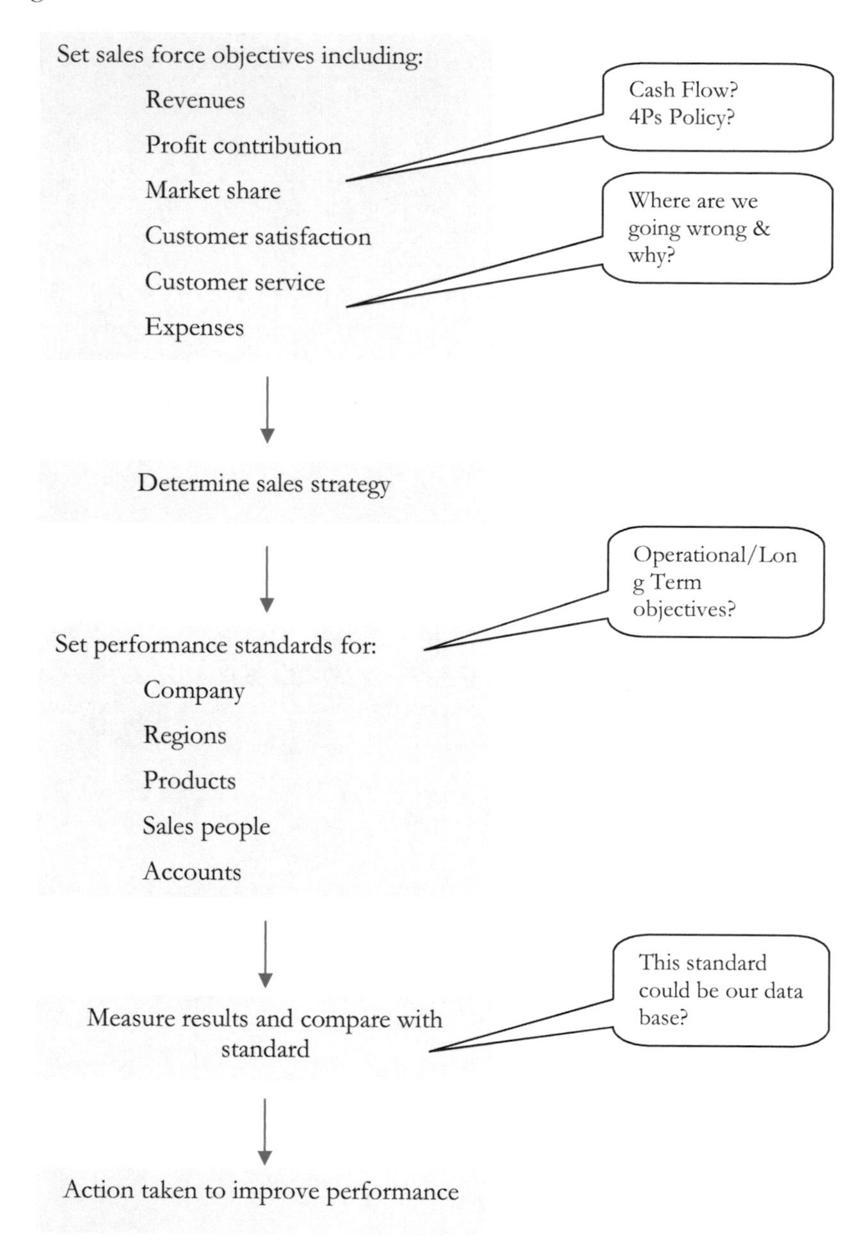

Fig 55 – adapted sales force evaluation model Jobber & Lancaster ((2003. P.444).

For this example however the latter can be considered appropriate only to our existing product mix and product enhancement/deletion strategies arising out of our operational and long term objectives.

No where have we stated how we will review and evaluate innovation in respect of our integrated refrigerant product. Existing data is encouraging albeit it was obtained using a micro market audit. Eugene has expressed a wish to examine the overall philosophy envisaged with the innovation to confirm or otherwise resource needs and availability.
Arising out of global warming and the impact upon the ozone layer/material disposal the IVS innovation considers integrating the refrigerant needs of the majority market. Here they have looked at the number of refrigerators and freezers sold on average each year. They have assessed the percentage of those who have purchased separate fridge/freezer units as well as those who would also purchase a domestic (and light industrial/office environment) cooling system.

The innovation considers the removal of individual compressor/cooling systems in favour of one system that would link all of the devices where some form of heat exchange associated with cooling will take place. One would then purchase a specific product (refrigerator for instance) and plug it into a suitable receptacle.
NPD/Innovation process selection criteria is available, however its usage is highly dependent upon how we would wish to assess the basic factors used to derive such a process. Davidson (1997) is seen to suggest a specific R&D profile map but excludes any reference within it to individual category activity as a percentage of the total R&D process (Hooley et al (2004)). Furthermore, it will be seen that text book reference is more suited to those companies (or assumes that?) who operate already using marketing principles to create their strategic objectives.
For this example, our NPD process considers the application of a text book paradigm but with added criteria in block diagram format as shown in figure 56.

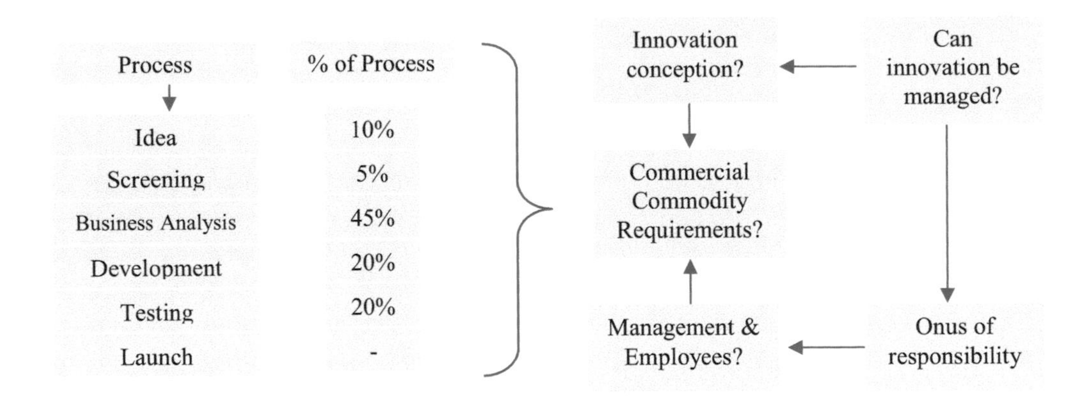

Fig 56 Modified NPD/Innovation Process Hooley et al (2004)

The creation of this particular algorithm would seem to be suitable in that it defines a specific procedure for undertaking NPD one should adhere to. Similarly, technology be it of a high scientific content or otherwise is either available or can be readily adapted (boundary definition known) to meet the creation of that which is to be developed, tested and launched.

Generating ideas and their screening for conversion into our *business plan* would suggest problems at the outset for those companies not familiar with marketing practise. For instance, an idea already exists for NPD/Innovation consideration for this example. But what is it that will impact the most on the methods employed to screen this idea and the consequent impact upon the *business plan*? There are several suggestions for consideration some of which are available from several authors:

Author/s	Research observations/conclusions
Maile, C.A; Bialik, D. M (1988)	Idea generation is of paramount importance during the NPD process. Concludes that less idea generation creates fewer products.
Meldrum, M.J (1994).	• Does the product generate unique marketing demands? • Can paradigms or models be developed for specific areas of attention? • How can high technology be distinguished in terms of strategic marketing, i.e. what themes that are important occur on a regular basis" • Lack of marketing and R&D interface within company organisations appears to be a common theme. • High technology products need to have distinguishing attributes. • Few products consist of a single technology. This can provide attributes in terms of differentiation." • Risks perceived by clients in terms of usage can relate more to the packaging rather than the function it is intended for. • Need to establish factors that affect relationship between suppliers, products and markets. He suggests that "this is closer to the heart of what marketing is all about". • Supports this author's comments given in that there are technology life cycles as well as product life cycles. • Initial product development is high in profile and is likely to yield low returns. Later small developments can produce

	high returns.
Beard, C; Easingwood, C (1992).	• Good high technology marketing is "technology push" based. Competitive advantage or superior product function must be identified. Don't just identify basics based on an Ansoff analysis. • Product improvement processes are determined by a market pull strategy. • Product quality and customer services are gaining competitive advantage.
Dunn, D. H; White. S.D, (1999).	• Any NPD must acknowledge needs arising out of the typical sales channels associated with the company, e.g. - Vertical Marketing Strategies – sells to one or several specific industries. - Horizontal Marketing Strategies – produces for a broad spectrum. • Any NPD must concentrate on added value performance, e.g. - Specialised Solutions - Customised Solutions - Value Solutions - Packaged Solutions Here this author would suggest that these four factors could then be subdivided into further factors for consideration (Baker 2000), namely: - Need elements - Need intensity - Need stability - Need diffusion Source: Rothwell (1983)

Benckenstein, M; Bloch. B (1994).	• Needs to be a balance between technology push and market pull. • Market Positioning - is technology the fundamental success factor? • System Technology is observed to have the ability to provide technology leadership. • "Market Entry" and "Timing" options will always have a high level of importance within any NPD process employed.

Hooley et al (2004)	• Company's functional capabilities and resources must be borne in mind
	• Technology must be exploited in a novel/new way
	• The intent should be to maintain technological leadership
	• Consider strategic role associated with:
	  - Product type – is it new?
	  - Product type – is it new to the world?
	  - Product type – is it a new product line?

Kotler, P; et al (2002)	• Have produced an expanded NPD model that suggests new product strategy prior to idea generation. However, this *business plan* would argue that this analysis already forms part of marketing plan analysis • They go on to include concept development and testing and marketing strategy prior to business analysis which would most probably be associated with any R&D that attracts high expenditure
Cahill, D.J; Warshawsky R.M (1993).	Supports Meldrum in respect of the importance of R&D/Marketing interface. One of the writers occasionally assembles microcomputers but has difficulty programming the VCR. A task his learning-disabled son finds easy to achieve.

Table 10.2

At first glance the process given may appear complex. It is the dissemination of the process activity using the table provided that has the potential to simplify all activities owing to basic ontological knowledge as applied by R&D (or engineering) and the department responsible for managing marketing activities. Here the latter is seen to have a significant impact upon any R&D activity and its management. The block diagram depicted as figure 57 below is used to define the relevant influencing factors.

Prudence would suggest that the process identified above would not form part of *business plan* presentation format but that the data is included within the appendices attached to the plan. In reality R&D as a function of our performance review and evaluation can be depicted as a single entity together with our adapted sales force evaluation for our next sub chapter presentation needs. This is given as Fig 54.1

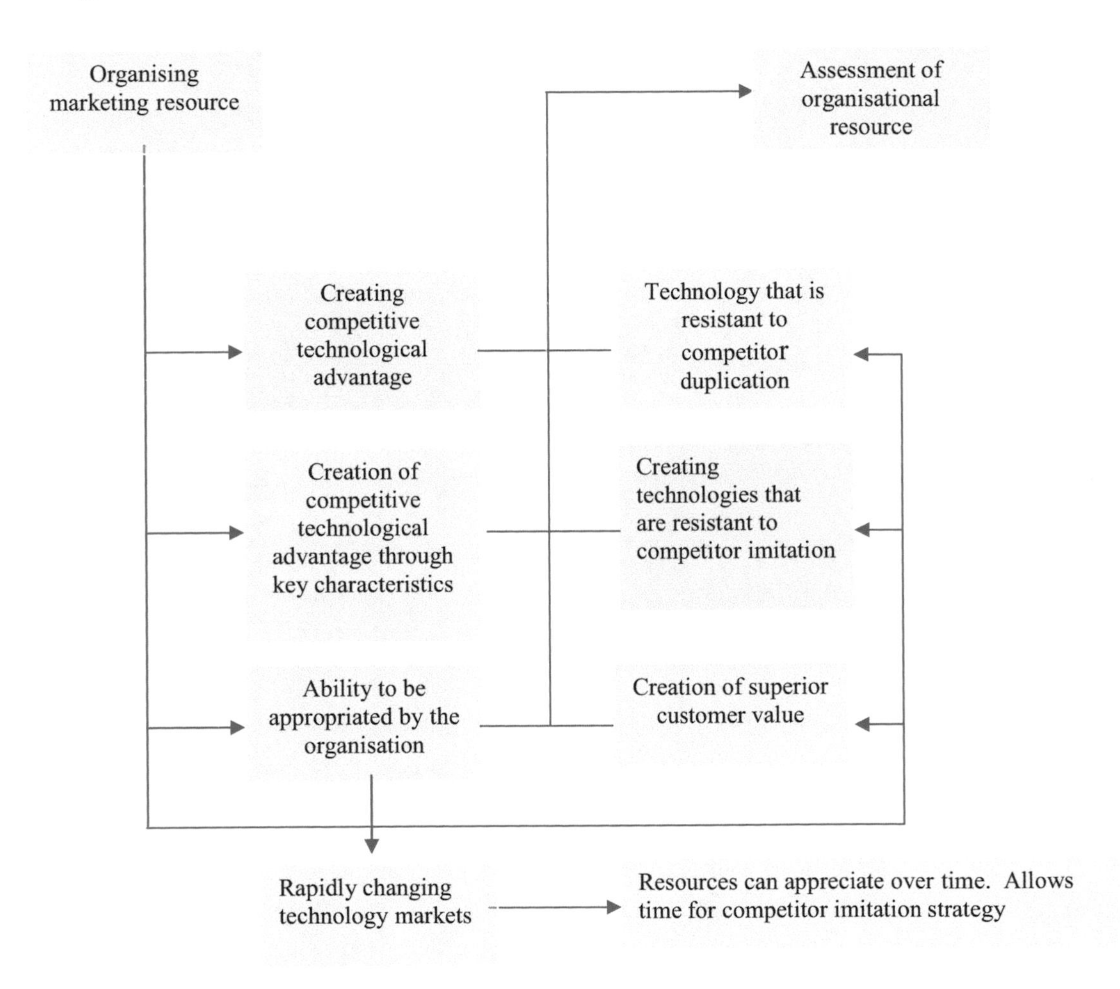

Fig 57 Understanding Marketing Resource adapted from Hooley et al (2004)

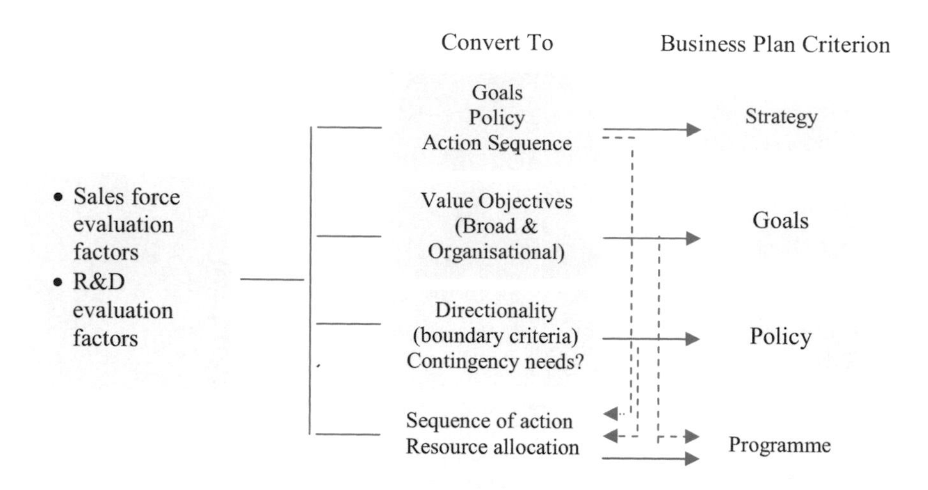

Fig 54.1 Model of Market performance assessment adapted from Baker (2000) & Donaldson (1997)

The second part of our performance audit considers the financial status/accomplishment criteria we wish to associate with the *business plan*. Here we must be aware of the need to assess the latter criteria by way of how the *business plan* will have a positive impact upon the ratios used to measure performance. Put simply, in what way will the management reporting be undertaken to reflect changes arising out of the marketing plan.
We have observed that marketing is seen to influence the asset base (tangible, intangible and investment) and certain performance ratios for a given company. However, marketing is unlikely to have a direct impact upon stock turnover or debtor ratios for instance. These would be determined more by the sales process employed and/or management effectiveness. Ergo we again have an argument for incorporating a two tier reporting system the reasons for which would be identical to that stated at the beginning of this example. But should we employ two accounting systems? Common sense would suggest that only one system is used. Clearly, one that is able to extract data from which performance revue can be considered and evaluated. We must therefore revert back to our analysis of how we must convert the data generated by the data base that may be in use.

For instance any *programme* used as part of the *business plan* may be seen to impact upon stock control or the purchase of specific components stored as work in process. Thus our reporting format should be based upon *business plan* objectives rather than the provision of an accounting ratio. A simple analogy could be:-

> "The asset base has increased in value by 100k (%) which has resulted in 100man hours less per month but has not increased gross margin owing to a reduction in projected sales."

Here the emphasis is placed upon the impact a certain marketing activity has in relation to a financial ratio rather than stating the current/new value of the asset base. Similarly, a further sentence could be added where the same impact would reduce the sales process duration thus reducing quoted delivery periods. Thus we are able to present the following:-

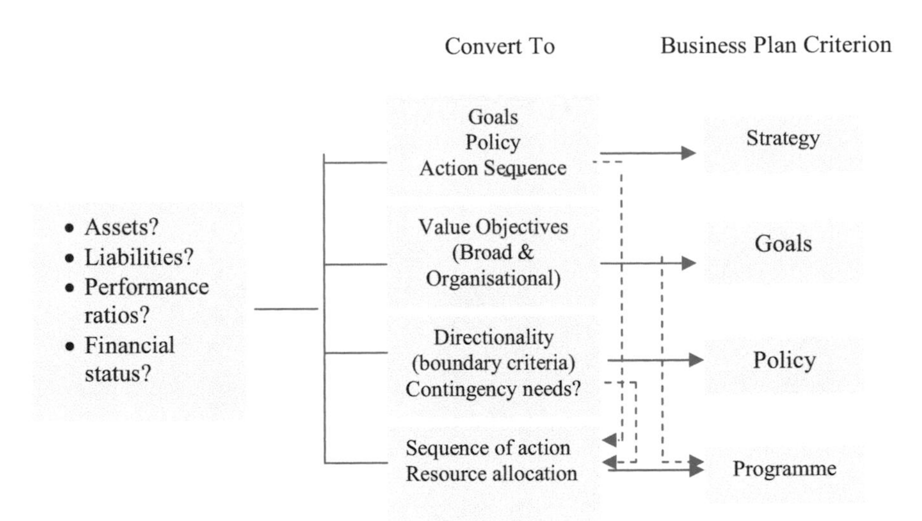

Fig 54.2 Model of Financial performance assessment adapted from Baker (2000) & Donaldson (1997)

Our report on attitudes should be applied in a similar vane. This being said, for this particular example it could be argued that we could extend *where are we leading?* and *where are we following?* with *where are we going?*.

Experience would suggest a greater level of *business plan* familiarity and its intent with senior and executive management than those employed at the process level. It is also suggested that this same level of familiarity is such that management presume of others the need of a specific action and its corresponding result on business performance.

This suggestion it could be said can apply both within an internal organisation as well as the market segments. It follows that our primary concern with our *business plan* attitudes audit should relate to the methods employed for its diffusion both internally and externally and where allowed its intent. Secondary and any tertiary issues would be those associated with *how*, *why* and *when* as well as what it is we would expect. We can then report on *where are we leading?* and *where are we following?*

Research would suggest that internal marketing as well as integrated marketing would to a large extent provide us with factors for review and evaluation when setting diffusion objectives. Several suggestions are put forward; table 10.3 refers.

Internal Marketing	
Author/s	Comment
Hooley, G; et al (2004)	• Management change requirements and why? • Focus on buyer needs in different industries (formal & informal) • Prevention of technology/customer orientation clash • Product management considerations
Piercy, N.F (2003)	• Strategic change in company arising out of market led theories, assertions & suggestions • Technological performance and the potential for change
Integrated Marketing	
Davidson, H (1997)	• Rapidly changing technology is rapidly expanding markets which in turn creates integrated marketing barriers • Technology application in terms of customer benefits still create rapidly expanding markets but with less barriers • Technology perception for what it is will not create rapidly expanding markets
Piercy, N.F (2003)	• Acknowledges the importance of the process of going to market and its impact upon company departmental requirements and R&D/Marketing departmental relationships
McDonald, M (2003)	• Customer relationship management developments are driving different trends in multi channel communications strategies
Kotler, P; et al (2002)	• Supports McDonald in respect of the changing communications environment • Message content must reflect company strength. • Company strength can be measured in terms of strategy, technology, processes and people

Table 10.3

We must also consider the psychological association of any diffusion techniques used; suggestions for which are given within fig 54.2A.

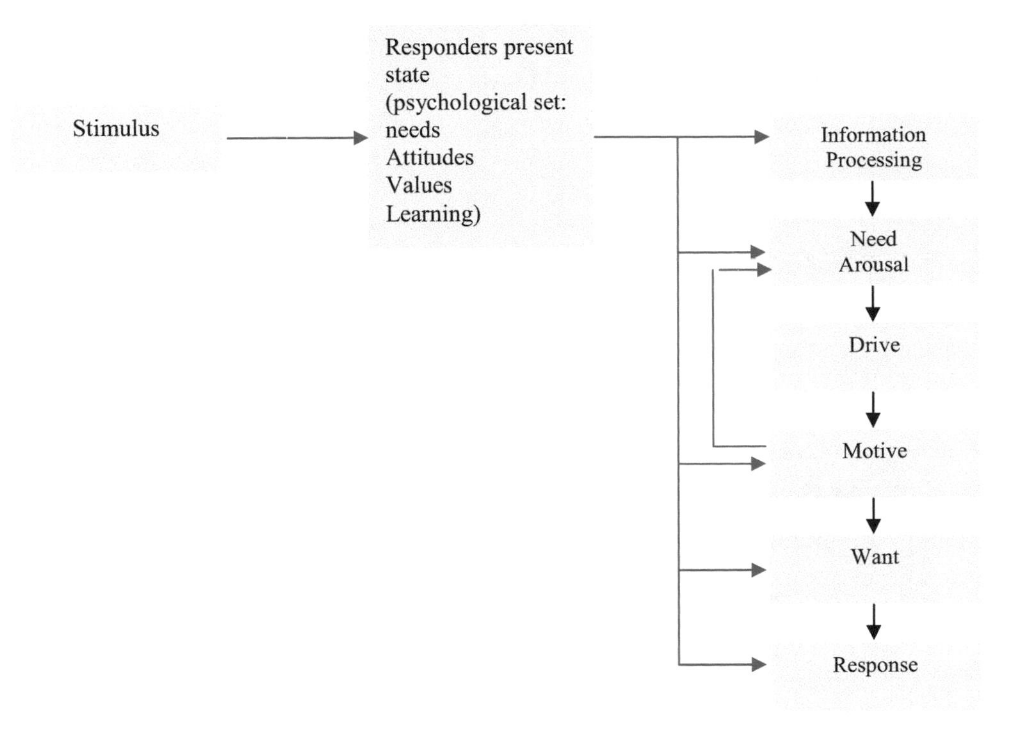

Fig 54.2A Source: Mahatoo, W.H (1989) Motives must be Differentiated from Needs, Drives, Wants: Strategy Implication. European Journal of Marketing, 23 (3).

Primarily for use as a research tool figure 54.2A can be seen to create a greater level of information accuracy/factual feedback. Our attitudes audit could therefore be summarised as given in Fig 54.3.

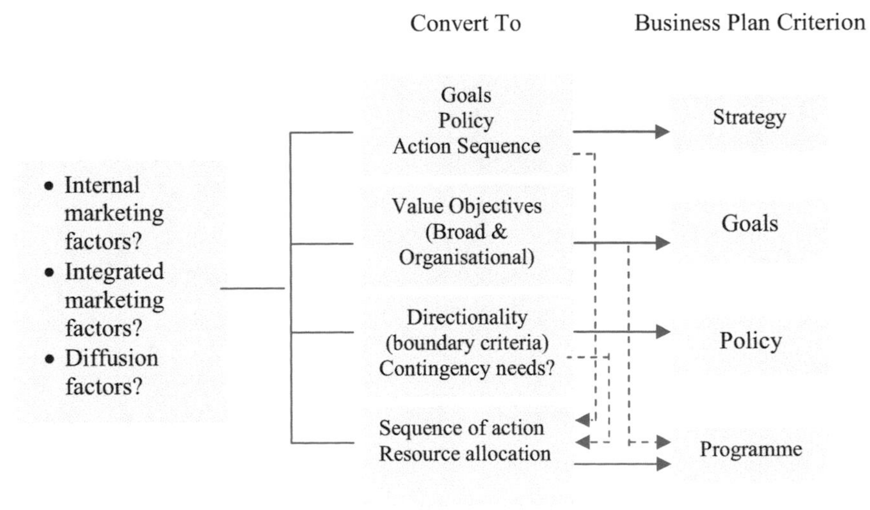

Fig 54.3 Model of Attitudes performance assessment adapted from Baker (2000) & Donaldson (1997)

It will be observed that for this particular sub chapter the primary input requirements are dependent upon all of the departments that the company is structured upon. This may contrast somewhat to the former sub chapter where primary input data is department specific.
The main chapter reference to the necessary items for inclusion within any strategy audit is to a large extent self explanatory. Barriers to simple and effective performance review and evaluation criteria are potentially those where the success of the *business plan* reflects only upon its association with the marketing plan. Yes the strategies employed have allowed us to achieve our objectives but it may have had a negative impact upon the sales process and/or the systems in force to review and evaluate sales process performance.

If we again review the constituent parts of figure 51 given above does it not follow that the marketing plan should have priority over the sales process? In that our resource allocation is seen to impact upon sales volume and market share. An argument would therefore exist that any adaptation of any sales process would be that required to enable it to deliver the marketing strategy employed. The *business plan* must seek to establish these important criteria.

The fact that our *business plan* takes into consideration long term strategy should provide a sufficient argument in this regard. To this end operational activities will always be subject to alteration to incorporate changes arising out of long term objective strategic requirements.

Thus we should now have the basic elements of the thought process to be employed to structure a performance review and evaluation audit. If we analyse the proposals put forward within the model of internal examination we must format the report in a manner that will allow us to respond to the typical factors put forward but in relation to market related conditions.
This would be the impact created by the strategy on the market, market viewpoint (Davidson) and its operational and projected long term performance or behaviour. As a result we can summarise the needs of this sub chapter; Fig 54.4 refers.

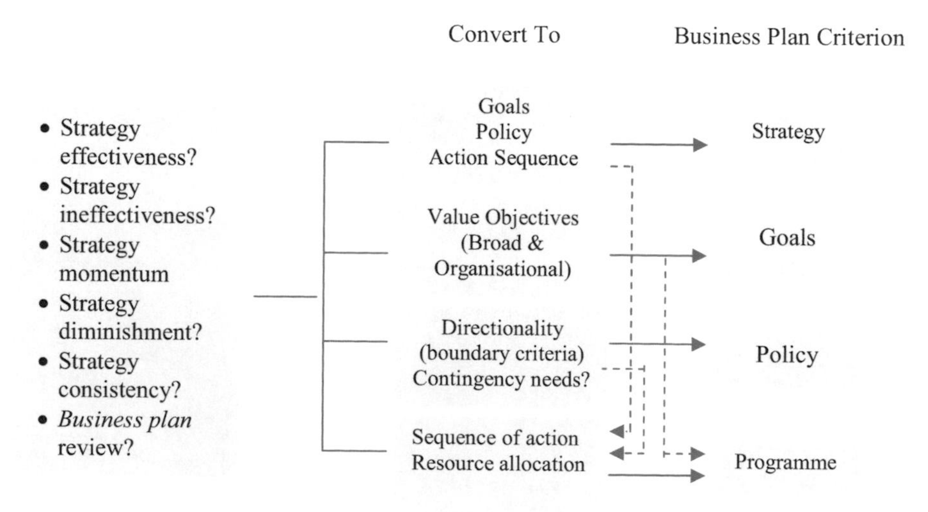

Fig 54.4 Model of Strategy performance assessment adapted from Baker (2000) & Donaldson (1997)

Our final sub chapter has the potential to give rise to a range of alternative methods of description. Irrespective of the nomenclature employed our execution audit is all about the effectiveness of strategy implementation. First to market is a simple example as would be lower costs to achieve the same end to an aim as a competitor.
Factors that contribute to our assessment of quality of execution are readily available for the assessment of NPD or innovation/invention. But in what way do we propose how to measure the quality of our execution for this *business plan*? This is made more difficult owing to the fact that IVS does not operate within a commodity based high volume market where quality of execution could be measured in terms of market leadership but by a large percentage. Factors for consideration are given as:-

Author	Measurement Factors
Davidson, H (1997)	• Suggest that effective execution is born out of the brand image created by the company • Brand image he argues are determined more by the key assets and competences within the company, typical factors are:- + High quality + Efficient production + Strong R&D + Low cost operation + High service levels + Strong supply chain + Effective selling + Low selling cost + Performance record + Financial strength + Integrity
Kotler, P; et al (2002)	• Suggests that sellers must identify the factors that buyers use to determine the effectiveness of their performance. They warn that different buyers will have different assessment factors.
McDonald, M (2003)	• Does not provide any sub chapter supporting criteria but includes evaluation using PhD research criteria in the form of a specific test (Smith)
Baker, M (2000)	• Argues that all determinant factors (marketing factors) may be fit

	for purpose or better but that it is the management skills that may detract from the quality of execution
Piercy, N (2002)	• Expands upon that provided by Baker by way of "execution skills of management" and the detracting strategy implementation techniques potential of" management by assumption", "Structural Contradictions", "Empty Promises" and "Bunny Marketing". Here the latter implies no clear strategy but a profusion of plans

Table 10.4

Given the above it could be argued that measurement of quality of execution could imply a long drawn out process if we are to measure the impact of our overall strategy. This would suggest that quality of execution should be measured using incremental factors. R&D for example and its *quality of execution*, which it could be said may be linked to our *performance* audit.

Furthermore we would need to analyse our *knowledge* base and how we are able to convert the data available. A sales individual may report that a certain client is highly delighted with our R&D programme but that the client would prefer an improvement in the development delivery programme. The programme would to all intents and purposes be dependent upon management skills to not only implement the R&D strategy but also the management of its completion in accordance with that suggested by Baker. Thus quality of execution could be improved if we convert the data into a format that would have a positive impact upon the supplier and the customer.
Here it is possible that additional finance and/or human competence may be required to meet the needs of the latter. This in turn raises the question of is the client prepared to assist with the financing of the R&D? Moreover, what impact will the R&D have upon the customer's own processes?
Here again the analysis provided would lend further support for the need of customer relationship management (CRM) techniques to be considered by a given company. For the example provided we will observe that it was a sales person reporting on quality of execution. The information could have been derived from a customer's engineer. Both are aware of the importance but neither has the overall say in strategic policy. Clearly, any such discussions would need to take place between more respective senior company individuals. Similarly, if we are to measure quality of execution using incremental factors does this not also support the need for CRM?

We can summarise our execution performance review and evaluation example analysis in several ways, some of which it will be noted have a probable crucial impact upon an organisation not fully observed by our text book authors.

- For those companies where the owners or corporate management team operate on a "hands on" basis information has been provided that will allow them to better understand business management techniques.

- For those companies where the owners or corporate management wish to operate on a "hands off" basis it will allow the identification of training that must be implemented where such skills may not currently exist. Congruently, management training in this regard will allow these same individuals to identify the training needs of support staff where such skills may not currently exist. Indeed, an additional chapter could be added to our business plan where skills training requirements are listed.

- To provided the basis of job functionality for those companies who are seeking to employ a senior or middle manager.

This is represented as shown in Fig 54.5

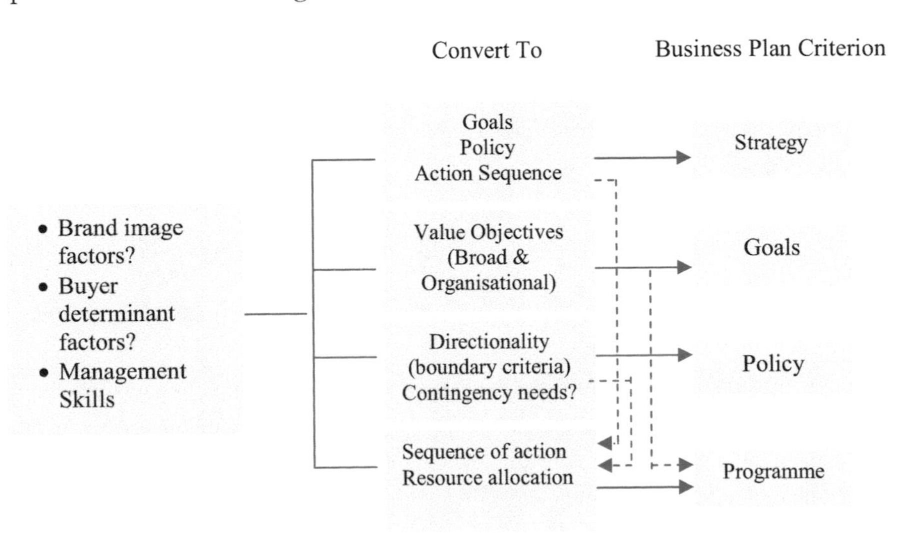

Fig 54.5 Model of Execution performance assessment adapted from Baker (2000) & Donaldson (1997)

Our final sub chapter and indeed *business plan* sub chapter is used to provide a point of reference. Substantial data may have been provided, albeit in a chronological methodological manner and there is a need to allow access to how it was structured and if need be a chronological method of modification. The more detailed marketing plan provided within the main chapter given above will allow a more direct approach to factors that may influence a change in *business plan* requirements.

For instance any changes arising out of the strategy audit where positioning techniques or methods of differentiation may need to be revisited. Before consideration is given to the implementation of any plan, be it marketing or otherwise it would be ideal if any barriers to such a plan could be observed. The plan given (Fig 51) analyses the potential barriers for a technological marketing plan.

However, as stated previously it contains too many variables for research consideration. McDonald, M (2003, P.52) the author of this plan also acknowledges that too much detail is provided. To this end, he suggests that the plan be used as a way of revisiting what it is that forms the basis of the strategic decisions made by management. Piercy, N.F (2003, P.582) uses supporting assertions by way of suggesting that a

> "good plan" will concentrate on actionable, and achievable considerations and be capable of being implemented.

Perhaps the most definitive need for this individual marketing plan requirement is that produced by McDonald, M (2003, P.52). Here he states;

> "Its major function is to determine where the company is, where it wants to go and how is it going to get there"

This author would wish to share the same sentiments put forward by Piercy, N.F (1995, P.4), here he states;

> "One of the most frequent, and certainly one of the most voluble pressures on management in all types of organizations is to focus on customer needs – to be customer led, to be market orientated, to care for the customers and so on".

How to Prepare and Implement an Effective Business Plan

Conclusions

In a world dominated by the internet and a rush to manufacture in low cost countries there is a tendency to forget about what people really want these days and how best you as business owners and managers can fulfil their needs. More so to ensure that these same needs are provided where favourable trading or vacuity conditions exist.

This environment or method of undertaking business seems to forget about how to set your business aside from others. Here there exists an argument to not only consider the cheapest prices available but also in what way will customers wish to use your product rather than others?
What is it that makes them aware of the factors that differentiate your business in a way that they are prepared to pay for what it is you provide feeling extremely satisfied with the purchase? Indeed, to perhaps undertake repeat purchases?

The world is forever changing and there is a need to change with the times. These same changes impact upon people needs and thus there must be a process that has the ability to analyse what if any adjustments or transformation is required to sustain or enhance business performance. Marketing as a science, it is hoped will allow readers to gauge if there is a need to change their perspective or to change the perspective of any potential customers or a mix of both.

It will be seen that the *Business Plan* put forward provides the foundation for a sound if not secure financial basis for any business. Five statements were made at the outset in relation to the benefits associated with incorporating a formal methodological process. These are summarised in Table 11.1

Methodological Analysis	Comment
To use text book "marketing planning" and "marketing strategy" techniques suggested by leading authors but to adapt them accordingly in an iterative manner.	Text book and research journals provide most if not all of the necessary demonstrative models and informative text required to produce a *business plan.* There is a need however to translate necessary syntax associated with academic needs into business operating philosophy and sales delivery and reporting techniques as given hereof.
To then structure the marketing plan into a format where its synopsis is used for "*business plan*" presentation techniques* but where its expanded format is used to provide corporate strategy guidance associated with business planning, implementation, measurement and control management.	Corporate strategy guidance is made available in a manner that allows business owners and managers to gauge more effectively their strategic aspirations or objectives in respect of current assets. Equally, business governance is made less tangible thus prompting "virtual" management techniques.

To use a financial summary as the point at which any marketing technique employed can be seen to impact upon. Perhaps more importantly, to provide a degree of flexibility for the structure of the financial summary and it's potential to vary in relation to market fluctuations.	Financial limitations or otherwise are made more flexible in respect of variable identification and the boundaries that they operate within in relation to maximising opportunities and risk limitation. Perhaps more importantly, they are provided in a manner that makes them more easily recognisable in respect of interdepartmental impact upon the business.
To structure a *business plan* that reflects its impact across a given market environment and the business as a whole including the activities of all individual company departments.	Allows businesses and all of their employees to acknowledge their competence and its impact within a given environment and the corresponding environmental accreditation associated with the services provided (Positioning & Differentiation).
Last but not least is the need to create revenue evaluation techniques using a chronological and iterative process in respect of available or planned competence resource that will maximise our opportunities and limit our risks.	Cash flow projections are structured using a more accurate reporting base in respect of opportunity maximisation and risk limitation.
Marketing Impact Upon the Business	Comment
It will allow the corporate management team a simpler method of business strategy identification and its subsequent management for all business types.	Creates a more cohesive approach to individual (micro) management operating techniques. In that, the competence base irrespective of departmental activity all operate using the same fundamental methodological process. Perhaps more importantly it provides business owners and senior managers an insight to the differences of managing the marketing and sales process requirement in relation to *business plan* management.
It will create an easier method of generating quality enquiries for small to medium organisations that do not employ any sales persons.	Many SME's may not be able to afford sales individuals. Here they will observe how they can more easily align their resources to specific market activity (segmentation, differentiation &

	positioning) in a way that will maximise their opportunities and minimise their risks.
Last but not least is the impact that put forward has in respect of company resource training needs. What skills levels does the company currently employ? What additional skill levels are needed and in what way will they benefit the business activity?	Many training courses exist using an eclectic array of why they should be used. Here business owners are able to audit their existing business activities to gauge what training may be required to maximise opportunities or to minimise risks or a mix of both.

Table 11.1

There is also a summary of Table 1.1 "Technological Impact" (P.24) to be considered. In particular its relationship to the overwhelming supporting criteria for the *business plan* example provided and its association with the identification of an additional directional research base where more text book supporting data is required, table 11.2 refers.

Technological Impact	Research Criteria
Technological Impact	Research Criteria
Technological reasons for marketing & society response criteria	Here it is argued that macro environmental factors associated with a given industry together with hierarchal factors (Governments etc) are considered in respect of our 4Ps.
Technological reasons that impact upon the creation of customer value	Market research clarification techniques that must be employed.
Technological Impact upon market demands	SWOT analysis and planning strategy techniques to be employed.
Technological reasons for market planning	Corporate objective prompting and analysis factors
Technological Impact upon opportunities and threats	Resource allocation and practicability in relation to planning strategy and P&L
Technological Impact upon the environment	SWOT analysis and planning strategy techniques to be employed. Competitor analysis, differentiation and positioning techniques that must be observed/implemented. P&L and cash flow

	(long term) analysis?
Technological Impact upon the marketing environment	Using technology in a manner that benefits society not just the company the produces it – Corporate Objectives Vs Environmental Factors.
Technological Impact upon consumer buying behaviour	Market research clarification techniques that must be employed. SWOT analysis and planning strategy techniques to be employed. P&L and Cash Flow impact?
Technological Impact upon marketing strategy	Business Plan construction techniques?
Technological Impact upon marketing strategy – resources	Balancing operational and long term strategies in relation to practicable P&L analysis factors.
Technological Impact upon "product"	"Directional" policies must be assessed in respect of all 15 factors given.
Technological Impact upon portfolio summary	Planning strategy must be clear and balanced in respect of "PEST" factors, Resource Base and required ROCE in relation to Operational & Long Term Strategic demands.
Technological Impact upon market segmentation	Segmentation techniques will directly impact upon P&L analysis but also very much in respect of business governance.
Technological Impact upon forecasting	The "Knowledge" base used and its method of reporting in relation to interdepartmental activity but also its overlap and impact with other departments.
Technological Impact upon distribution and sales policy	Sales processes need to be assessed in relation to marketing and business plan application.

Table 11.2

These tabulated conclusions by way of their allusion to "balancing" resource with competitive environmental factors would suggest some method of iterative financial reasoning to arrive at an acceptable P&L statement. It is suggested therefore that some form of data base be constructed using Fig 51 (Page 235) as the point of data input origin. One can then amend a given factor and observe its impact upon the P&L.

By adopting the suggestions put forward for business management and business process management we have created a more simple overall structural solution. This will to a large extent limit the "supermarket shopper" type management syndrome often found in large organisations. Here the analogy refers to those individuals with shopping trolleys who stop to check out their needs and block the aisles for other shoppers.

Reference was made earlier to the fact that Ferrari cars are not built in China. While this may not be acceptable to most as analogy let us consider another scenario. What if all manufacturers of a given type of product for the same market segment transfer their manufacturing activity to the same low cost manufacturing region? They all now have a low "direct cost" cost of sales basis. So in what way will they be able to differentiate their product or business activity from the competitors? Clearly, any lower price differentiation strategy for the same functional use will no longer exist.

It must also be stressed that globalisation to a large extent is a way of life for the larger conglomerates. It represent a relatively quick method of ensuring competitiveness within a given market segment without any radical change in product design or functionality. However, it is a proven fact particularly from a technological standpoint that smaller companies can co-exist within the same market segments as these large conglomerates. They are known to operate much closer to the market and are thus able to match their offering in terms of market needs.

Globalisation would seem to be an ideal scenario when considering commodity based marketing activities. Clearly, if seeking to purchase a 60W bulb for our garage we will seek the lowest price unit available to us. Furthermore, if we own a Ferrari would we be upset if the head lamp bulbs were not made in Italy. What we must guard against is a world where "Sales" based organisations prevails. Evidence, will show that the larger market intermediary or segment users have now evolved where the buying process has transformed the product selection process in a commodity based selection process. Put simply, product needs are based upon market usage at that time without any consideration for future applications. Clearly a dangerous situation for us all where monopolistic market segments is seen to impact upon us all. Is there any wonder why green house gases and the hole in our ozone layer remains a regular news subject?

Given the above we can conclude that there are many benefits for adopting a marketing plan as the basis of a *Business Plan*. These are those benefits that impact not only upon the company but across the macro environment as a whole.

Bibliography

Baker, M, J. (2000). Marketing Strategy and Management (3rd Edition). Basingstoke: PALGRAVE.

Beard, C & Easingwood, C. (1992). Sources of Competitive Advantage in the Marketing of Technology-intensive Products and Processes. *European Journal of Marketing*, 26 (12), 6-18.

Benkenstein, M & Bloch, B. (1994). Strategic Marketing Management in High-tech Industries: A Stock-taking. *Market Intelligence & Planning*, 12 (1), 15-21.

Brooksbank, R (1999). The theory and practise of marketing planning in the smaller business. *Marketing Intelligence & Planning*, 17 (2), 78 – 90.

Brooksbank, R; et al. (1999). Marketing in medium-sized manufacturing firms. The state-of-the-art in Britain, 1987-92. *European Journal of Marketing*, 33 (1/2), 103-120.

Cahill, M.J & Warshawsky, R.M. (1993). The Marketing Concept: A Forgotten Aid For Marketing High Technology Products. *Journal Of Consumer Marketing*, 10 (1), 17-22.

Chelsom,, J. M. (1998). Getting equipped for the twenty-first century. *Logistics Information management*, 11 (2), 80-88.

Christopher, M. (1970). The Existential Customer. *European Journal of Marketing*. 4 (3), 160 - 164

Coates, N & Robinson, H. (1995). Making industrial new product development market led. *Market Intelligence & Planning*, 13 (6), 12-15.

Crick, D & Jones, M. (1999). Design and innovation strategies within "successful" high-tech firms. *Market Intelligence & Planning*, 17 (9), 161-168.

Czuchry, A. J & Yasin, M. N. (1999). The three "Is" of effective marketing of technical innovation: a framework for implementation. *Marketing Intelligence & Planning*, 17 (5), 240 247.

Darden, W. R; et al. (1989). The Impact of Logistics on the Demand for Mature Industrial Products. *European Journal of Marketing*, 23 (2), 47 – 57.

Davidson, H. (1997). *Even More Offensive Marketing*. London: Penguin Books.

Dhanani, S; et al. (1997). Marketing Practices of UK high technology firms. *Logistics Information Management*, 10 (4), 160-166.

Dixon, D,F & Wilkinson, I.F. (1984). An Alternative Paradigm for Marketing Theory. *European Journal of Marketing*, 23 (8), 59 – 69.

Dunn, D; et al. (1999). Segmenting high-tech markets: a value added taxonomy. *Market Intelligence & Planning*, 17 (4), 186-191.

Gardner, D.M; et al. (1999). A contingency approach to marketing high technology products. *European Journal of Marketing*, 34 (9/10), 1053-1077.

Grantham, L.M. (1997). The validity of the product life cycle in the high-tech industry. *Marketing Intelligence & Planning*, 15 (1), 4 – 10.

Hart, S.J. (1989). Product Deletion and the Effects of Strategy. *European Journal of Marketing*, 23 (10), 6-17.

Hooley, G; et al. (2004). Marketing Strategy and Competitive Positioning (3rd Edition). Oxford: Pearson Education Limited.

Jobber, D & Lancaster, G. (2003). Selling and Sales Management (6th Edition). Essex: Pearson Education Limited.

Kotler, P; et al. (2002). Principles of Marketing (3rd European Edition). Essex: Pearson Education Limited.

Mahattoo, W. H. (1989). Motives must be Differentiated from Needs, Drives, Wants: Strategy Implications. *European Journal of Marketing*, 23 (3), (pp 29-36)

Maile, C. A & Bialik, D. M. (1988). An Extended Model For New Product Selection. *European Journal of Marketing*, 23 (7), 53-59.

McDonald, M. (2003). Marketing Plans: How To Prepare Them, How To Use Them (fifth Edition). London: Butterworth-Heinemann.

Meldrum, M.J. (1994). Marketing high-tech products the emerging themes. *European Journal of Marketing*, 29 (10), 45-58.

Meredith, J. (1993). Theory Building through Conceptual Models. *International Journal of Operations & Production Management*, 13 (5), 3 – 11.

Meredith, J; et al. (1989). Alternative Research Paradigms in Operations. *Journal Of Operations Management*, 8 (4), 297 – 326.

Moncrief, W.C; Cravens, D.W. (1999). Technology and the changing marketing world. *Market Intelligence & Planning*, 17 (7), 329-332.

Nijssen, E.J; Lieshout, K.F.M. (1995). Awareness, use and effectiveness of models and methods for new product development. *European Journal of Marketing*, 29 (10), 27 - 44.

Palmer, P.J & Williams, D.J. (2000). An analysis of technology trends within the electronics industry. *Microelectronics International*, 17 (1), 13 – 16.

Piercy, N.F.(1995). What do you do to get customer focus in an organisation? *Marketing Intelligence & Planning*, 13 (6), 4 – 11.

Piercy, N.F. (2003). Market-Led Strategic Change (3rd Edition). Oxford: Elesevier Butterworth-Heinemann.

Saunders, M; et al. (2003). Research Methods For Business Students (3rd Edition). Essex: Pearson Education Limited.

Simkin, L & Dibb, S. (1998). Prioritising target markets. *Marketing Intelligence & Planning*, 16 (7), 407 - 417.

Slowikowski, S; Jarratt, D.G. (1997). The impact of culture on the adoption of high technology products. *Market Intelligence & Planning*, 15 (2), 97-105.

Thomas, K.W & Tymon, W.G. (1982). Necessary Properties of Relevant Research: Lessons from recent Criticisms of the Organizational Sciences[1]. *Academy of Management Review*, 7 (3), 345 – 352.

Wang, Q & Montaguti, E. (2002). The R&D – marketing interface and new product entry strategy. *Marketing Intelligence & Planning*, 20 (2), 82 – 85.

Watson, A; et al. (2002). Consumer attitudes to utility products: a consumer behaviour perspective. *Marketing Intelligence & Planning*, 20 (7), 394 – 404.

Zineldin, M. (2000). Beyond relationship marketing: technologicalship marketing. *Marketing Intelligence & Planning*, 18 (1), 9-23.

Printed in the United States
110926LV00002B

* 9 7 8 1 4 3 4 3 2 1 0 9 1 *